Embedded Systems – A Hardware-Software
Co-Design Approach

Bashir I. Morshed

Embedded Systems – A Hardware-Software Co-Design Approach

Unleash the Power of Arduino!

 Springer

Bashir I. Morshed
Department of Computer Science
Texas Tech University
Lubbock, TX, USA

ISBN 978-3-030-66807-5   ISBN 978-3-030-66808-2   (eBook)
https://doi.org/10.1007/978-3-030-66808-2

This Springer imprint is published by the registered company Springer Nature Switzerland AG
The registered company address is: Gewerbestrasse 11, 6330 Cham, Switzerland

This book is dedicated to my parents
Dr. Monjur Morshed Mahmud
and
Suraiya Begum
who inspired me and believed in me
and always loved me and supported me.
—Bashir I. Morshed, Ph.D.

Preface

Rapid progress of embedded systems (ES), with a myriad of products including smart devices, Internet-of-Things (IoTs), wearables, and robotics, has led to the newer and smarter future. Learning about these requires understanding both hardware and software, a concept known as hardware–software co-design. These hardware design and software algorithms need to process the continuously collected data in real time for actionable decision-makings with high degree of accuracy and reliability. Resolving these challenges require approaches that integrate: hardware–software co-design paradigm, resource constrained data collection and processing, on-chip processing, power saving, memory-management, and wireless data transfer schemes, edge and cloud computing, machine learning and deep learning classifiers, autonomous and semi-autonomous feedbacks with control systems in a meaningful and timely basis. This book attempts to provide fundamentals to ES by covering both of these aspects. As a platform for specific codes, Arduino Uno is utilized throughout the text.

Lubbock, TX, USA   Bashir I. Morshed

Contents

Chapter 1
Introduction

The scientists of today think deeply instead of clearly. One must be sane to think clearly, but one can think deeply and be quite insane.—Nikola Tesla, one of the greatest minds.

1.1 What Is Embedded System?

Embedded system (ES) is a term coined to represent systems that are designed specifically to be deployed and operated independently in a predetermined or an intelligent fashion while being completely embedded within a unlike system or environment. In general, ES represents electronic computing devices embedded within embodying system. Peter Marwedel has defined ES (Dortmund definition) as follows: "Embedded systems are information processing systems embedded into enclosing products" [1].

In contrast, general-purpose computing represents traditional computers (e.g., desktop, laptop, and computer server). Computing elements other than general-purpose computing falls into ES category. In essence, nearly any computing system other than general-purpose computers can be considered as ES. However, there are some technical issues with such generalizations that are still unwarranted. For instance, smartphones are becoming general purpose in usage, although historically mobile phones were strictly ES. It is generally accepted that ES are electronic devices containing embedded computing elements that are typically realized with microcontrollers. Traditionally, these systems were developed with a certain application in mind, but this design approach is also evolving. Lewes [2] has given the following four characteristics for ES:

1. An electronic device that incorporates a processing unit such as a microcontroller.
2. It contains complete hardware and software needed to fully operate the electronic device.
3. It is designed and can operate independently without human intervention.
4. The device is optimized to handle a few specific tasks only (as per design).

As mentioned, ES will typically be comprised of a microcontroller with analog/ digital input/output ports, sensors, actuators, one or more display, liquid crystal

© Springer Nature Switzerland AG 2021
B. I. Morshed, *Embedded Systems – A Hardware-Software Co-Design Approach*,
https://doi.org/10.1007/978-3-030-66808-2_1

Fig. 1.1 Typical hardware
composition of an ES

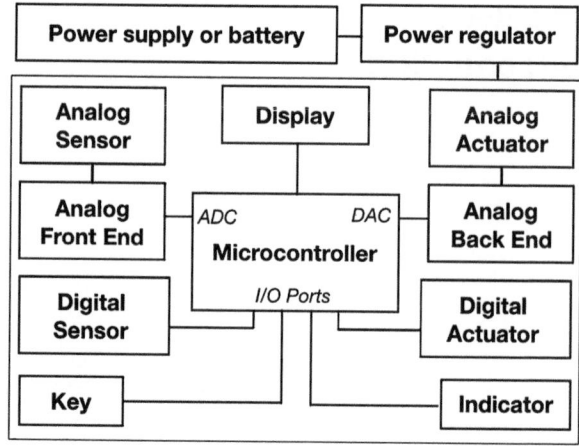

display (LCD), and light emitting diode (LED) indicator, and a power module [3]. A
generic hardware composition of a typical ES is shown in Fig. 1.1. Microcontrollers
have digital input/output (I/O) ports for communicating to peripheral devices. Most
microcontrollers also have integrated analog to digital converter (ADC) to collect
analog signal processed by analog front-end electronics from analog sensors. Some
microcontrollers also have digital to analog converter (DAC) to output analog
signals for analog back-end electronic circuit (e.g., driver circuit) that can operate
analog actuators. Sometimes ES can have displays such as LCD or graphical
monitors. In addition, LED or other indicators are used for status and user input
(e.g., Keypad) can be connected to the digital I/O ports. A battery typically powers
an ES through a power regulator for portability, but ES can also be powered from
wall power supply, if appropriate.

In general, ES can be characterized as having three distinct features:

(a) *Function specific:* Most ES can perform only a few specific tasks. For instance, a
calculator can only compute numbers as entered by the user, and a microwave
oven can only heat food for the set time with specified heat setting. At the
deployment, the function cannot be changed (e.g., calculator cannot heat food or
microwave oven cannot perform computation!).

(b) *Tightly constrained:* Most ES designs are tightly constrained due to the limited
resource availability. For instance, battery operated ES must consume very small
amount of power to ensure long operation between recharge, as well as circuits
must be designed to operate within the voltage supplied by the battery (unless a
voltage booster is utilized). Other constraints can be low-cost of materials, small
physical size, light-weight, limited memory availability, etc.

(c) *Real-time reactive operation:* Many embedded systems need to respond to
human inputs in real-time in a reactive manner for action and feedback. For
example, when we press the elevator key, we expect an indicator to notify us in
real-time that our input has been accounted for. The reactive nature can also be in
response to an environmental stimulus. For instance, a room heater can start

operating when the temperature of the room is below a certain set value. The ES designer must take these into account such that the required computation must be performed, and an action is assigned in real-time without any significant delay. In some cases, these constraints are critical, for example a drone must respond very quickly to sudden air gust by changing the power to different rotors so that the stability of the drone is not compromised otherwise it might lead to catastrophe for the system or to the embodying environment.

Designing ES is, thus, complex and the designer must be aware of all design factors and operate within the constrained design space. This becomes even more complex and challenging for some ES whose testing in actual setting is impractical such as design of Mars Rover, dangerous such as fire alarm system, or expensive such as drone control module. ES designer must consider these factors at design time:

(a) *Efficient:* Designer must consider design alternatives for an efficient and cost-effective ES while considering small size, light-weight, and long run-time without human intervention.
(b) *Reliable:* ES must be designed in such a way that the system works correctly all the time. Designers should consider failsafe aspects for even in the worst possible scenario.
(c) Maintainable: ES should be maintainable periodically so that partial failure of the system does not lead to complete failure and replacement of the system. Maintainability might also allow upgrading part of the system without incurring cost of replacement of the complete system.
(d) *Safe:* A lot of ES operates with human in contact and needs to be safe to avoid harm to human and surrounding environment.
(e) *Secure:* ES is expected to be secure to enable users trust. Security and privacy concerns must be addressed. For example, communication must be confidential and authentic.
(f) *Real-time:* As mentioned, most ES operate in environment where real-time response is required as the system is constrained by stimuli from the environment and objectives.

As an example of ES used in daily life, consider a home air conditioning (providing both heating and cooling) system (Fig. 1.2). For simplicity, we will only consider temperature as the input stimuli to the system. This core air conditioning system will have a temperature sensor as the input sensor and a keypad for user to select preferred settings. An onboard LCD display can show the current settings, current temperature, and the system status. The ES controls a heating unit if the setting is in Heat Mode, and it controls a cooling unit if the setting is in Cool Mode. For Heat Mode, when the temperature is below the set value, the heater will turn on, whereas for Cool Mode, when the temperature is above the set value, the cooler will turn on.

ES, a cyber entity, frequently interacts with a physical environment as in this example. Hence, these types of systems are sometimes referred as cyber-physical

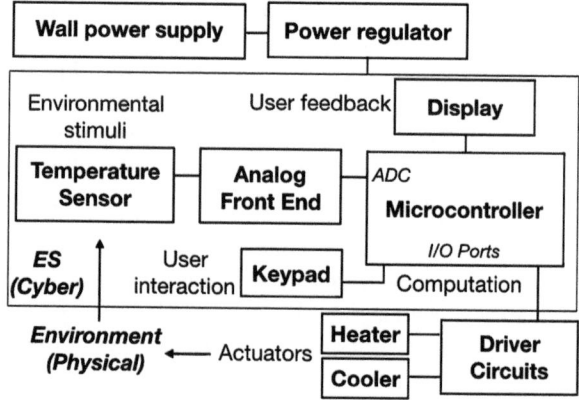

Fig. 1.2 A basic air conditioning system hardware blocks as an example of an ES

system (CPS) [4]. CPS is also typically a feedback system: the physical system causes the cyber system (viz. ES) to monitor and reactively provide stimulus to the physical system via actuators, which might cause the physical system to modify or adjust its physical state such that the CPS reaches to a stable operating point. For instance, considering air conditioning system shown in Fig. 1.2, if the temperature of the room air (a physical system) is different from the set temperature, the ES (a cyber system) will cause the heater or the cooler to be activated (depending on the user settings) such that the temperature reaches to the set temperature (the stable operating point). Typically, ES feedbacks are achieved with simple control loops (e.g., loop subroutine in Arduino code), however newer ES with complex tasks (e.g., computer in self-driving car) needs advanced techniques such as interrupt-controlled system, multitasking and parallel processing, multi-threading, mutex, and scheduling. In this book, we will emphasize these advanced techniques for control using both hardware and software systems with Arduino Uno as the prototyping platform, although the concepts discussed are applicable to other platforms as well (e.g., MSP430 from Texas Instruments, PIC microcontrollers from Microchip Technology, PSoC microcontroller from Cypress Semiconductors).

1.2 History of Embedded Systems

Apollo Guidance Computer is considered as one of the earliest systems with all components of ES. It was developed at MIT Instrumentation Laboratory by Charles Stark Draper in the early 1960s [5]. The system contained thousands of integrated circuit (IC) chips (4100 for Block I and 2800 for Block II) with NOR gates performing the computations with word length of 16-bits. The system had an integrated interface with display and keyboard (DSKY). The software (instruction set) was custom developed assembly program with 3 bits for opcode and 12 bits for address. The Block I design had 12 kilowords of fixed memory (later updated to

24 kilowords), and the Block II design had 32 kilowords of fixed memory and 4 kilowords of erasable memory. The system used a 2.048 MHz crystal clock.

The ES technology has come quite a long way since then. It has progressed drastically in higher computation capability and lower cost with the advent of microcontrollers, higher memory capacity, and increased clock speed. In the latter half of 1960s, complex ICs were developed for high-volume use and companies like Intel was formed for commercial production of ICs. The trend in technological growth tends to follow Moore's Law predicted in 1965 by Intel co-founder Gordon Moore. The law states that IC transistor capacity will double roughly every 18 months, and the trend is observed for the past several decades.

By the early 1980s, microcontrollers have integrated input/output system components as well as memory into the same chip as the processor. The size and weight of newer devices are becoming smaller with the same or better functionalities, while the comfort and unobtrusiveness are increasing at a rapid pace [6]. For instance, the first electrocardiography (ECG/EKG) device was a desk-sized equipment where the user had to insert hands and a leg in saline water baths to capture the signals. The equipment had reduced to an ambulatory bench-top system with the invention of vacuum tubes. When discrete semiconductors (viz. transistor) were invented, the ECG/EKG device size reduced to small portable devices. With advances in integrated circuits, currently smart wearable devices can collect this signal. It is anticipated that these systems will become as small as e-skin circuits or epidermal electronics with nanotechnologies such as graphene, organic electronics, and flexible printed circuits.

1.3 Embedded System Market

The advantage of microcontrollers is a comparatively low-cost and size compared to a large number of separate input/output ports, memory chips, and some other components (e.g., internal clock). As a result, ES can be built with little design and development time with low non-recursive expense (NRE). This is attractive for business as they can bring products to market with within short time and with little research and development (R&D) expense.

Consumer demand for ES market is growing as ES provides comfort, functionality, and reliability to consumers. In a 2009 survey, it was found that ES market size (49%) has exceeded traditional computer market (19% of PC workstations and 20% of peripherals) [7]. At that time, the ES revenue was the most in Americas (over $35B), but the growth rate in Asia-Pacific was the highest (~22%) followed by Europe (~13%). Just the global mobile entertainment industry itself was worth about $32B with an average annual revenue growth of 28%. Another industry affected by ES was remote home health monitoring, which generated $225 M revenue in 2011, up from $70 M in 2006, according to Parks Associates. Newer industries such as the identity and access management (IAM) market share in Australia and New Zealand (ANZ) reached about $190 M by 2012. The ES hardware market exceeded $100B in

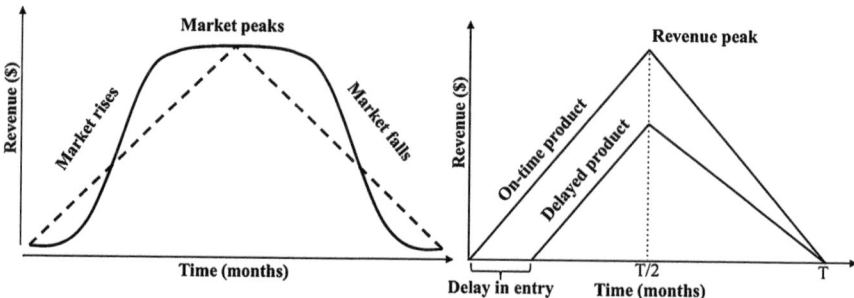

Fig. 1.3 (Left) A typical market window curve for ES. Dotted line shows triangular approximation model. (Right) A triangular approximation model depicting revenue loss due to delay in entry for an ES product

2013 and still growing at a fast rate [8]. Accessing the Internet via mobile devices jumped by 82% in US and by 49% in Europe between 2007 and 2008. In fact, we are still observing an exponential growth with projections like a trillion devices world-wide by 2030.

From the business prospective, ES is lucrative as time-to-market can be drastically reduced with very small NRE cost. Time-to-market indicates the time required to develop a product from concept development to the point that it can be sold to the customers. For ES, the time-to-market can be as small as 6 months! As most ES product market window, period of time during which the product would have sales, is limited, small time-to-market is desirable. To analyze this, let us consider a typical product life cycle. Figure 1.3(left) shows a simplified market window curve where market demand starts to grow leading to increase of revenue. At some point, market saturates and the product revenue peaks. After some time, market demand starts to fall, due to reasons beyond the product itself, such as access to newer technology, change in market need, or shift in perspectives. The dotted line shows a triangular approximation of the curve for ease of analysis. The area under the curve determines the total revenue generated from this product.

Let us analyze the triangular approximation model shown in Fig. 1.3(right). If the total product life is T months, the revenue peaks at $T/2$ months. The area of the triangle defines the total revenue (R), which can be calculated with the formula:

$$R = \frac{TP}{2} \tag{1.1}$$

where P is the peak revenue. If a product is delayed by D months, assuming the same rate of market penetration rate as the on-time product, the peak revenue of the delayed project is:

$$P' = \frac{2P}{T}\left(\frac{T}{2} - D\right) \tag{1.2}$$

Hence, the total revenue of the delayed product can be calculated using the formula:

$$R' = \frac{P}{T}\left(\frac{T}{2} - D\right)(T - D) \tag{1.3}$$

Potential revenue loss can be defined as the difference between the total revenue generated by the on-time product and the total revenue generated by the delayed product. This loss is the area difference between the two triangles. Percentage revenue lost by the product delay can be computed by this formula:

$$\%\text{Revenue Loss} = \frac{R - R'}{R} \times 100\% = \frac{3TD - 2D^2}{T^2} \times 100\% \tag{1.4}$$

Example 1.1 For a product life cycle of 52 months, calculate the percentage revenue lost for a product delayed by 4 months.

Solution Product life cycle, $T = 52$ months, and product delay, $D = 4$ months. Using the formula (1.4), we get % Loss of Revenue $= 21.89\% \approx 22\%$.

Example 1.2 Consider the same product life cycle of 52 months, but another product that is delayed by 10 months. What is the percentage loss of revenue for this case?

Solution Product life cycle, $T = 52$ months, and product delay, $D = 10$ months. Using the formula (1.4), we get % Loss of Revenue $= 50\%$.

This product delayed by less than one fifth of the product life cycle has lost half of the potential revenue. Thus, product delay can be deadly for companies.

Another important market factor to consider for ES products is the per-product cost. It is defined by the cost to produce each unit by considering NRE cost. Unit cost is the monetary cost of manufacturing each copy of the product excluding NRE cost, which is the one-time monetary cost of design and development of the product. Considering the total number of units produced over the product lifetime, the total cost can be determined by adding NRE cost with the product of the unit cost and the number of total units produced. Hence, per-unit cost can be written as:

$$\text{Per unit cost} = \text{Unit cost} + \frac{\text{NRE cost}}{\text{Number of units}} \tag{1.5}$$

Example 1.3 If the NRE cost of a product is $2000 and unit cost is $100, what are the per-unit cost if the total number of units produced are (a) 10 units, (b) 1000 units.

Solution NRE cost $= \$2000$ and Unit cost $= \$100$.

(a) Per-unit cost = $100 + $2000/10 = $300.
(b) Per-unit cost = $100 + $2000/1000 = $102.

As seen in this example, amortizing NRE cost over a larger number of units leads to a lower per-unit cost. Typically, ES products end up to smaller production volumes, as there are many options, rapid changes of technologies, and smaller market windows (compared to technologies such as microprocessor).

1.4 Future Trends

Currently, billions of ES units are produced yearly versus millions of desktop units. It is projected that by 2030, there will be trillions of ES systems and edge nodes worldwide connected to billions of ES gateways to communicate with cloud in a massive interconnected system (Fig. 1.4). Tremendous growth and usage of ES is forecasted for future. Some of the most impactful emerging sectors that ES is predicted to significantly influence are discussed below.

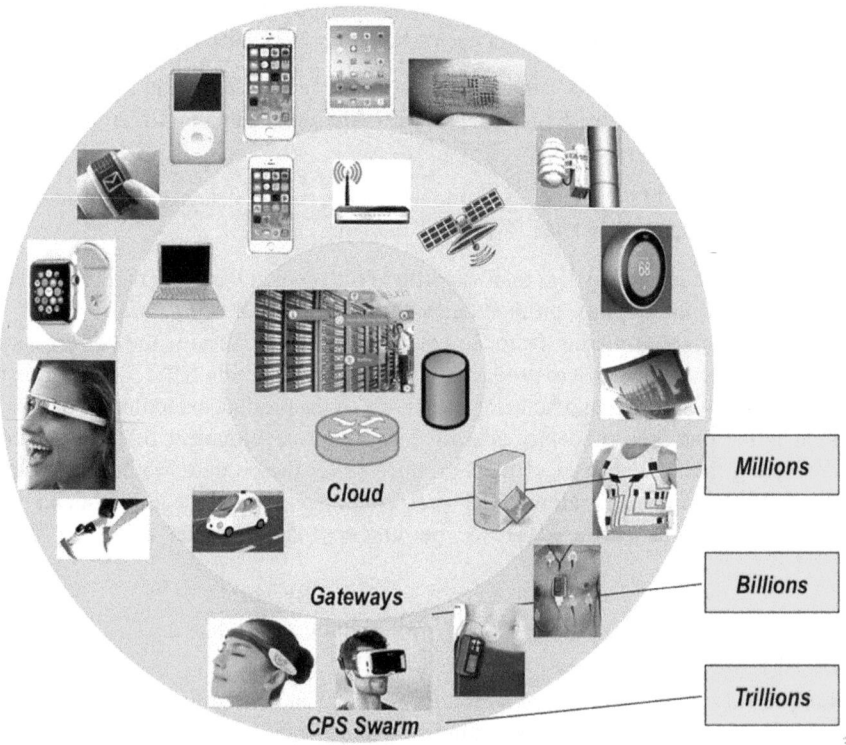

Fig. 1.4 Vision 2030 for ES forming a cyber-physical system (CPS) swarm

1.4.1   Internet of Things (IoT)

Internet of Things (IoT) are ES devices with internet connectivity that can be access or controlled remotely through Internet. IoT provides significant capability advantage to the users. For example, "Ring" devices allow the user to see and hear the person at the door via internet that allows user to proactively and accurately decide entrance allowance and security even away from home. IoTs will be centrally connected to cloud where high complexity computation can take place (e.g., deep learning). The computation performed at the central servers (viz. cloud computing) will have seamless communication for sensing/actuation at IoTs with small computation capability (viz. edge computing). Some computation can also be performed near edge such as router (viz. fog computing) to improve the system capability.

1.4.2   Wearables

Another type of ES device is such that they can be worn on the body. Examples of wearables are smart watches, fitness related gadgets, and smart glasses. Wearables can collect data of user activities as well as allow presentation of information in ways that were previously not possible. Future wearable can be smart tattoo, smart clothing, and smart electronic patch sensors. A distinction can be made for devices that can be worn such that the device do not move with respect to skin (such as epidermal electronics, electronic patch sensors) which can be termed as "Body-worn" devices, whereas other devices that are not fixed with respect to skin rather allowed to move while worn on the body (such as smart watches, smart phones) which can be termed as "Wearable" devices [9].

1.4.3   Electronic Medical Devices

A variety of ES can allow clinically related information to be collected from users or even take medically correct administrative procedures. Insulin pump system can monitor blood glucose continuously, and administer insulin as needed by the body in a closed loop control mechanism. Fully implanted pacemakers can monitor heart beat rhythms and provide a jolting stimulation if abnormalities are observed, thus saving users from life-threatening heart failure conditions. Many wearables can also serve as medical devices, e.g., electrocardiography (ECG/EKG) monitor. Electronic medical devices can be body-worn, minimally invasive, or fully implanted.

1.4.4 Artificial Implants, Limbs and Organs

Many artificial implants, limbs and organs have been developed using ES technology such as artificial arm, artificial leg, artificial heart, artificial retina, and artificial kidney. Medical prosthetics with electronic controls can allow disable patients to improve their quality of life. Artificial organs can be life-saving. Most artificial organs can also be classified as implantable medical devices.

1.4.5 Autonomous Vehicles

There is a significant push in recent years (latter half of 2010s) to develop cars that can drive themselves, known as "Autonomous Car" or "Self-Driving Car." Coupled with the possibility of electric vehicles, a promising future with drastically different mode of transportation is envisioned. Autonomous driving of many other vehicles is also being researched such as aircrafts, trains, and ships.

1.4.6 Smart Homes, Offices, and Buildings

Smart homes, offices, and buildings are residences or office spaces that have appliances, lightings, heating, air conditioning, computers, audio-visuals, security, and access systems that are capable of communicating with one another and can even be controlled remotely by users, automated by manual time schedule, or even connected to artificial intelligence and machine learning techniques to optimize energy and improve quality of life.

1.4.7 Cyber-Physical Systems (CPS)

CPS is integrations of computation, networking, and physical processes through embedded computers with the ability to control the physical process and consist of feedback loops where physical process affect the cyber component and vice versa. Most ES can be classified in these systems if they fulfill the criteria.

1.4.8 Robots and Humanoids

Science fiction stories and movies have long described about robots and humanoids. With the rapid advancements of ES, they are no longer fiction rather is becoming reality. Boston Dynamics has developed multiple forms of robots and humanoids including "Big-Dog." Humanoid "Sophia" developed by Hong-Kong based company Hanson Robotics. Sophia was activated on April 19, 2015. The human shaped humanoid is equipped with advanced artificial intelligence and can display more than 62 facial expressions. In November 2017, Sophia has been given citizenship by Saudi Arabia.

1.4.9   Drones

Drones are remotely controlled aerial vehicles than can fly or hover while collecting data, videos, or images. They are also termed as Unmanned Aerial Vehicle (UAV). Typical small sized drones are quad-copters, which can hover in air with relative ease. Usage of drone in rescue operation and warfare are emerging applications. Drones also found application in military operation such as surveillance or precision strikes. Recreational usage of drone is also increasing; however, concerns of privacy and safety need to be carefully examined.

1.4.10   Smart Power Grid

Smart energy monitor with sustainable electric power system is typically referred as smart power grid. Traditional electric power system (Power Grid) consisted of large power plants typically away from the user's locations connected through power transmission lines, substations, transformers, and distribution lines. In Smart Power Grid, the power can even be generated by the users using Solar panels or Wind Turbines and can be fed to the "Grid" to generate revenue for the users. This system includes a variety of operational and energy measures including smart meters, smart appliances, renewable energy resources, and artificial intelligence to optimize energy efficiency and reduce energy bill.

1.5   ES Design Metrics and Challenges

A design metric is a measurable feature of a system's implementation. The ES system designer must consider multiple competing design metrics. Some common design metrics are listed below.

(a) **Unit cost:** The cost of manufacturing is critical for successful product. Unit cost is the monetary expense for each copy of the system excluding non-recursive engineering (NRE) expense.

(b) **NRE cost:** This is the one-time expense to design and develop the system. A lower NRE cost makes good business sense.

(c) **Size of the device:** The physical space required by the system should be small. This is particularly important for wearables.

(d) **Performance:** The performance of the system required depends on the specific application. A real-time system must respond quickly enough to ensure timing guarantees, where execution time is very important. Some systems require high throughput while some other systems might require fast digital signal processing (DSP).

(e) **Power:** The amount of power consumed by the system is typically very important as these ES devices are operating 24/7. Small saving in power can end up saving a large amount of energy to run these systems. Power optimization is especially critical for wearable systems.

(f) **Flexibility:** It is interesting to conduct NRE cost of a product with flexibility of the ability to change the functionality of the system without suffering additional substantial NRE cost.

(g) **Time-to-prototype:** The time required to develop a working prototype should be as small as possible to make "go-no go" decision of the product as well as to improve functionality.

(h) **Time-to-market:** As time-to-prototype, a small time to a final product also gives an edge to the company in the market to capture early market share.

(i) **Maintainability:** Another important design consideration, which is typically ignored by designers, is maintainability. The system should have the ability to modify the system after its initial release. This instills confidence to the consumer and higher level of satisfaction.

Some other design factors to consider are accuracy, precision, safety, reliability, and redundancy. Optimizing design metrics is a key challenge of ES designers. While ES designers would like to construct an implementation with desired functionalities and optimal design metrics, some of these design metrics can be challenging to optimize, as there are competing metrics. For instance, a system with improved computational power might lead to higher power consumption and needed higher development time and NRE cost. Similarly, a smaller size might end up with sacrificing performance or increasing unit cost. ES designers must encounter the challenge and attempt to simultaneously optimize numerous design metrics with prudent tradeoff decisions.

Exercise

Problem 1.1: Using a space heater as an example of an embedded system, draw a hardware block diagram of this system. Other than essential components, the system contains a manual keypad entry, a forced air heater, option for oscillating $120°$ range, and a display to show the set temperature and the sensed ambient temperature.

Problem 1.2: For an automated cashier machine at a store, draw a hardware block diagram. Other than essential parts, this system has a barcode scanner, a weighing machine, a card reader, a cash collector, a cash change dispenser, a cash reloading mechanism, a green/red light signaling mechanism to indicate status, a secure back-end connectivity with a financial transaction machine via Internet, a camera, a diagnostic port, and a touch sensitive display.

Problem 1.3: For an embedded system product with 3 years life cycle, a delayed product was launched after 6 months compared to on-time products. Compute the percentage potential revenue loos for the delayed product using a triangular revenue model approximation.

Problem 1.4: A company is planning to launch a new embedded system product. They estimate the product will have 4 years of life cycle. To avoid more than a quarter potential revenue loss, what is the maximum delay they can afford? Use a triangular revenue model approximation.

Problem 1.5: For an embedded system product with a market life of 60 months, compare the revenues for an on-time product and a delayed by 10 months product. Assume the revenue peak is $100 K. Use the triangular approximation model for the market window. Also, determine the percentage loss of revenue for the delayed product.

Problem 1.6: For an embedded system product, the NRE cost and unit cost are the following for the four technologies:

Technology	NRE expense	Unit cost
Semi-custom VLSI	$200,000	$5
ASIC	$50,000	$10
Programmable FPGA	$15,000	$20
Microcontroller	$10,000	$15

(a) Calculate total per-unit cost for production volumes of 100, 1 k, 10 k, and 100 k units.
(b) Plot these data from (a) in a single graph with log scale for per-product cost and draw piecewise linear lines for each technology. Then, determine the best choice of technologies for these production volumes (100, 1 k, 10 k, and 100 k units) to achieve the lowest per-product cost. Also plot total cost for these product volumes in a separate log-log graph.
(c) From the per-product cost plot in the (b), estimate the range of production volumes for which each of these technologies is financially optimal.

(d) List 3 other considerations in addition to per-product cost that might affect the choice of technology.

Problem 1.7: If a product unit cost is $40, and NRE cost is $10,000, determine the minimum quantity of units to be produced to keep per-unit cost below $50.

References

1. P. Marwedel, Embedded System Design: Embedded Systems Foundations of Cyber-Physical Systems, 2nd edn, (Springer). (2011)
2. D.W. Lewis, Fundamental of Embedded Software with the ARM Cortex-M3, Pearson education, 2nd edn (2013)
3. R. Kamal, *Microcontrollers: Architecture, Programming, Interfacing and System Design* (Pearson Education, India, 2005)
4. E.A. Lee, S.A. Seshia, *Introduction to Embedded Systems: A Cyber-Physical Systems Approach*, 1st edn. (UC Berkley, 2012)
5. E.C. Hall, *Journey to the Moon: The History of the Apollo Guidance Computer* (AIAA, Reston, Virginia USA, 1996)
6. Y. Zheng et al., Unobtrusive sensing and wearable devices for health informatics. IEEE Trans Biomed Eng **61**(5), 1538–1554 (2014)
7. Chip Design Trends—ASIC prototyping survey, 2009
8. "Embedded Systems: Technologies and Markets" and "Future of Embedded Systems Technology". BCC Research Reports (IFT016C, G-229R)
9. S. Consul-Pacareu and B.I. Morshed, Design and Analysis of a Novel Wireless Resistive Analog Passive (WRAP) Sensor Technique, *IET Wireless Sensor Systems*, in press (2018)

Chapter 2
System Modeling

> All models are wrong, but some are useful.—George E. P.
> Box, one of the great statistical minds of the 20th century.

2.1 Modeling Needs

Before diving into designing hardware and software of an Embedded System
(ES) project, ES designers should consider important related design questions.
Examples of design questions are:

- What type of microcontroller is suitable for this project?
- What type of inputs and outputs are required?
- What are the sensors, actuators, and communication choices?
- What is the best partitioning of software and hardware?

A system model with abstraction is a good choice to start with. Modeling allows
ES designers, project managers, business officers, and customers to exchange ideas
early to ensure proper design choices are made. System modeling allows efficacious
communication among the stakeholders and allows strategic distribution of
responsibilities.

2.2 Specifications and Requirements

Specification refers to a method for defining the overall system objectives in a
quantitative manner. Specification allows designers to narrow down sub-system
and component choices, as well as developing constraints. Consider the room air
condition system shown in Fig. 2.1. What are the design questions to address of this
system? It provides an overall notion of inputs and outputs, but what information
should be transferred for proper operations of various blocks? How to decompose
the system for modular development process and hardware–software partitions?

© Springer Nature Switzerland AG 2021   15
B. I. Morshed, *Embedded Systems – A Hardware-Software Co-Design Approach*,
https://doi.org/10.1007/978-3-030-66808-2_2

Fig. 2.1 A simple ES room
heating-cooling system
conceptual diagram

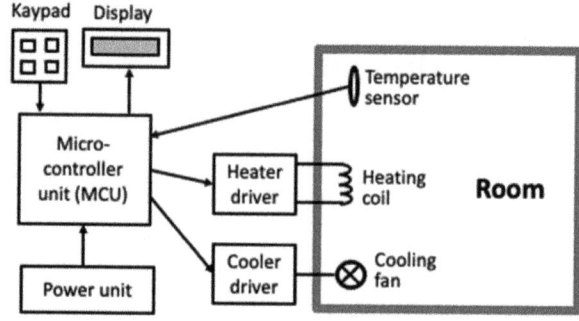

Fig. 2.1 A simple ES room heating-cooling system conceptual diagram

Requirements are prerequisites that are used as guidelines. Requirements are not absolute rules, rather allows flexibility. There are typically two sets of requirements: Minimum requirement and Recommended requirements. Minimum requirements must be met by the ES design; otherwise the design must be iterated and optimized until these requirements are met. On the contrary, recommended requirements are not critical to meet, but every effort should be made to meet these requirements as well. ES designers can develop multiple constraints for design space to meet these requirements. Constraints define the boundary space within which a designer must search the optimal solution to the problem. ES designs are typically constrained by many competing criteria. ES designers would like to optimize performance while keeping cost to a minimum but faces dilemma with schedule.

These specifications and requirements might lead to a stringent design space where optimal design might not be the best solution to the problem. Allowing some tolerance is a good design practice that takes advantage of extended design space. Tolerances can be provided as percentage of error or a range of minimum and maximum values.

As an example, let us reconsider the room air condition ES (Fig. 2.1). The system will turn ON a heater when the temperature of the room is low and turn ON a cooling fan when the temperature of the room is high. Specification can indicate a target room temperature, for example 72 °F. Requirement is that the air condition must maintain this temperature of the room during normal operating condition. However, if the operating condition is beyond normal range, for instance an extremely cold day or an extremely hot day, the ES might not be able to meet this requirement. Hence the minimum requirement should define what constitutes a normal operating condition. Recommended requirements are that the ES must attempt to its best to maintain the target room temperature. Constraints of this system can be set as maximum energy that the system can consume. The target room temperature must allow some tolerances; otherwise the system will alternatively turn on the heater or the fan for small fluctuation of the room temperature. For instance, a tolerance range of 71.5–72.5 °F will allow the system to operate more robustly.

2.3   Modeling Approaches

In the very early phases of some ES design project, only descriptions of the system under design (SUD) can be available in natural language. These descriptions can be misleading as interpretation of natural language might change with individual perspective or discipline. Hence, systems must be modeled using formal approaches with tools. These tools should be machine-readable so that abstraction and efficient communication can be possible. Version management is also important as the design progress through stages and gets refined. Another important aspect of these tools is analysis of dependency.

The process first starts with identifying the objects of a system. Identifying objects and describing them in a proper way is an important part of the process during object-oriented analysis. The objects should be identified with corresponding responsibilities, which represent the functions performed by the object. It is also important to identify the relationships of the objects and the requirements for their fulfillment. Objects can collaborate according to their intended association. It is a good practice to approach this modeling with the understanding that the design should be made in such a way so that it can be converted to executable using a modeling language. Typically, modeling language is implemented using object-oriented languages.

For system level models of ES, conceptual modeling is a good approach. A conceptual model is a model made of concepts and their relationships. These types of models describe the behavior of various objects or classes in the system and identify interoperability. Conceptual models are made of concepts and their relationships. The purpose of this type of modeling is to design the behavior of classes. In addition, these models can contain state machine diagrams that allow signal sending, signal receiving, loop, tests, composite states, and setting or using of variables. There are various approaches for modeling. Modeling can start from a very generic structure with only inputs and outputs connected to a black box (known as "Level 0") and can be decomposed progressively to more specifics ("Level 1," "Level 2," etc.). Modeling can also be done with objects. Modeling objects with state machines is called automata.

2.4   Unified Modeling Language (UML)

Unified Modeling Language (UML) is a state machine (TAU G2 state machines) based conceptual modeling approach. UML is a standard language for specifying, visualizing, constructing, and documenting artifacts of software systems [1]. UML was created and standardized by Object Management Group (OMG), a non-profit consortium of industry, academia, and government. UML 1.0 specification was drafted and proposed to OMG in January 1997. A major revision was conducted in July 2005 and UML 2.0 was released. In 2009, UML 2.2 standard was released.

OMG is continuously putting effort to make a truly industry standard. UML definitions are driven by consensus, rather than innovation. UML is different from the other common programming languages like C++, Java, Python, etc. as UML is a pictorial language with some coding. The purpose of UML is to develop a blueprint of the software to be made. In addition, UML allows shared development more streamlined and strategic.

UML is a standard methodology to organize and consolidate the object-oriented (OO) development. It also defines some general-purpose modeling language that all modelers can use. UML is simple to understand and use, thus UML diagrams are not only useful for developers but also for business users, common people and anyone interested to understand the system. Note that UML is not a precisely specified model of the computations. It is also not a precisely specified model of communication. Although UML was originally developed for software, it can be applied for non-software and hardware–software combination systems like typical embedded system projects.

UML users can be developers, testers, business people, analysts, and many more. UML uses the following language definition styles:

- Abstract syntax (meta-model)
- Static semantics (well formedness rules in OCL)
- Dynamic semantics (textual, in natural language)
- Concrete graphical syntax
- Usage notations

2.4.1 UML Basics

UML can be described as the successor of object-oriented analysis and design and is adopted for ES design methodology. UML can describe real-time systems. There are three major elements to UML programming:

1. UML building blocks,
2. UML rules that connect these building blocks, and
3. UML diagrams that define common mechanisms of the system.

UML uses fundamental principles such as abstraction, encapsulation, inheritance, and polymorphism. Different UML diagrams describe various facets of the model. Each diagram is basically a projection of the model. Thus, several diagrams of the same kind may coexist. Similarly, not all system will require all the diagram representation. The UML design is more fluid; hence strict syntax or structure is not used, rather the design reflects the thought process of the designer, which can be easily interpreted by other stakeholders. However, due to this nature of UML, incoherence among diagrams (same type or different types) can arise or realization might be impractical. These are examples of potential ill-formed models. The possibilities of ill-formed model exist in UML, as the coherence rules between different kinds of diagrams are not fully stated.

2.4.2   UML Definitions

In UML, an **object** contains both data and methods that control the data. The objects represent real-world entities. The **data** represents the state of the object. A **class** in UML describes an object. The class of an object can be considered a blueprint of the object. Classes form hierarchy to model the real-world system. The hierarchy is represented as inheritance of classes. The classes can also be associated in different manners as per the requirement. Like most other programming languages, **abstraction** represents the behavior of a real-world entity that is relevant to hierarchy. **Encapsulation** is the mechanism of binding the data together and hiding from outside world. **Inheritance** is the mechanism of making new classes from existing ones. **Polymorphism** is the mechanism for an object to exist in various forms.

2.4.3   UML Building Blocks

There are three main types of UML building blocks:

1. Things
2. Relationships
3. Diagrams

Things are the most important building blocks of UML. Things can be of four types:

(a) Structural things
(b) Behavioral things
(c) Grouping things
(d) Annotational things

Structural things define static parts of the model. They represent physical and conceptual elements. Examples of structural things are class, interface, collaboration, use case, component, and node. **Class** represents set of objects having similar responsibilities. **Interface** defines a set of operations that specify the responsibility of a class. **Collaboration** defines interaction between elements. **Use case** represents a set of actions performed by a system for a specific goal. **Component** describes physical part of a system. **Node** is a physical element that exists at run time. Representations of various structural things are shown in Fig. 2.2.

A class notation has a name provided in the top section. In the next section, attributes such as variable names and types with visibilities can be provided. Common visibility specifications are: Public, Private, and Protected. The

Fig. 2.2 Representations of various structural things

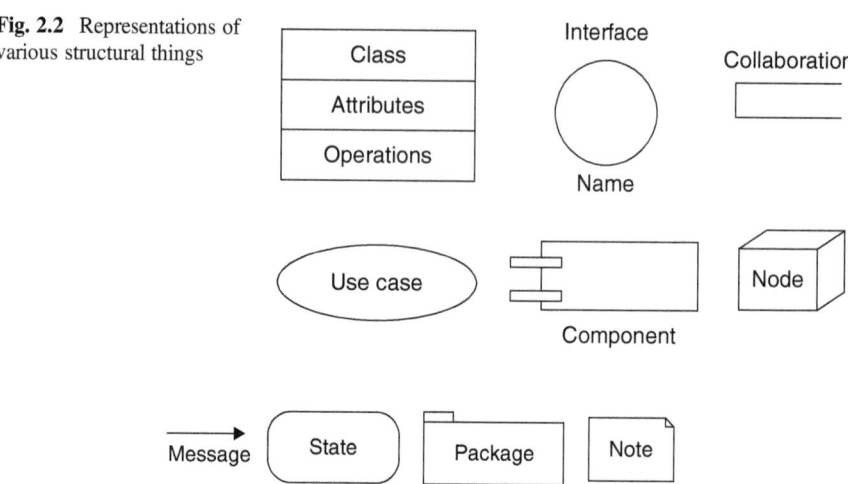

Fig. 2.3 Representations of (from left to right) Interaction (behavioral thing), State Machine (behavioral thing), Package (Grouping thing), and Note (Annotational thing)

corresponding symbols for these attributes are: "+", "-", and "#". The third section provides a list of operation of the class. They can be implemented using functions in programming code. Operations can also have visibility attributes. An extra section can be added (optional) that lists responsibilities. Comments are prefixed with "--" symbol.

The object, the actual implementation of a class, is represented in the same way as the class. The only difference is the name, which is underlined. An object can be considered as the instance of a class. The name of the interface notation is typically written below the symbol (circle). The interface defines functionality only, not implementation. Similarly, collaboration defines responsibilities, and represented with ellipse symbol. Generally, responsibilities are clustered together in a group. The use case notation captures high-level functionalities of a system, and also represented with ellipse symbol. The name of the use case is provided in the middle of the ellipse, and additional components can be used for clarification. Similarly, the component name and the node name also appear in the middle of the corresponding symbols.

A behavioral thing consists of the dynamic parts of UML models. One of the behavioral models is **interaction**, which is defined as a behavior that consists of a group of messages exchanged among elements to accomplish a specific task. **State Machine** (SM) is another behavioral thing that is useful when the state of an object in its life cycle is important. It defines the sequence of states an object goes through in response to events. Events are external factors responsible for state changes. Figure 2.3 shows representation of Interaction and State Machine.

Grouping things are a mechanism of a UML model to group elements together. For instance, **Package** is a grouping thing for gathering structural and behavioral things. The name of the package is written inside the package notation. Additional

Fig. 2.4 Representations of the relationships: (from left to right) Dependency, Association, Generalization, and Realization

components can also be used for clarification. A representation of Package is shown in Fig. 2.3.

Annotational things can be defined as a mechanism to capture comments, remarks, and descriptions of UML model elements. A **Note** is an annotational thing that is used to render comments, constraints, etc. within an UML element. The text inside a Note is only used for additional information and not part of the required programming. Figure 2.3 shows a representation of a Note notation.

Relationship is another important building block of UML. It depicts how elements are associated with each other and this association describes the functionality of an application. There are four kinds of relationships available in UML:

(a) Dependency,
(b) Association,
(c) Generalization,
(d) Realization.

Dependency is a relationship between two things in which change in one element also affects the other one. It is denoted by dashed line with a terminating arrow, with the name written above the line. The object at the arrow end is the Independent object, while the object on the other end is the Dependent. **Association** is a set of links that connects elements of an UML model that are related based on association. The notation is dashed line with arrow on both ends. The name of the association appears in the middle of the line. It also describes how many objects are taking part in that relationship, by using a number notation above the arrow. Star "*" symbol is used to denote many. This notation of multiple objects in a relation is called Multiplicity. For instance, many homes might be connected to one power grid information broadcasting system. **Generalization** is a relationship that connects a specialized element with a generalized element. It describes inheritance relationship n with the world of objects. **Realization** can be defined as a relationship in which two elements are connected. One element describes some responsibilities, which are not implemented while the other one implements them. This type of relationship exists in case of interfaces. Figure 2.4 shows representations of various relationships.

There are two types of UML diagrams:

(a) Structural diagrams
(b) Behavioral diagrams

Examples of structural diagrams are:

(i) Class diagram

   (ii) Object diagram
  (iii) Component diagram
  (iv) Profile diagram
   (v) Package diagram
  (vi) Composite structure diagram
 (vii) Deployment diagram

Examples of behavioral diagrams are:

   (i) State machine diagram
  (ii) Communication diagram
 (iii) Use case diagram
 (iv) Sequence diagram
  (v) Activity diagram
 (vi) Timing diagram
(vii) Interaction overview diagram

Composite structure diagram, timing diagram, and interaction diagram did not exist in UML 1.x, rather added in UML 2.0. Figure 2.5 shows some examples of various UML diagrams. One of the very important diagrams is the Use case diagram that uses an "actor" symbol to represent an internal or external entity that interacts with the system. Use cases describe applications of the system under design. In fact, the center of UML diagram is the use case view which connects all elements of the system. A use case represents the functionality of the system. The other perspectives are connected with the use case.

Design of an embedded system using UML typically consists of classes, interfaces and collaboration. UML provides class diagram, object diagram to support this. Implementation defines the components assembled together to make a complete physical system. UML component diagram is used to support implementation perspective. Process defines the flow of the system. So, the same elements as used in Design are also used to support this perspective. Deployment represents the physical nodes of the system that forms the hardware. UML deployment diagram is used to support this perspective. The structural modeling captures the static features of the system. This provides the framework for the system, but it does not describe the dynamic behavior of the system. Behavioral modeling provides description of the interactions in the system among the structural diagrams and shows the dynamic nature of the system. Behavior modeling can be describing (partial, declarative) type consisting of use cases, interactions (such as sequence diagrams, collaborations), high-level activity diagrams, and logical constraints. However, they might not provide any direct transformation into implementation. Another approach of behavior modeling is perspective (imperative, detailed), which consists of state machines (or StateChart) and actions. This is closer to implementation and describes the design coherently. Architectural modeling conveys the overall framework of the system and contains both structural and behavioral elements of the system.

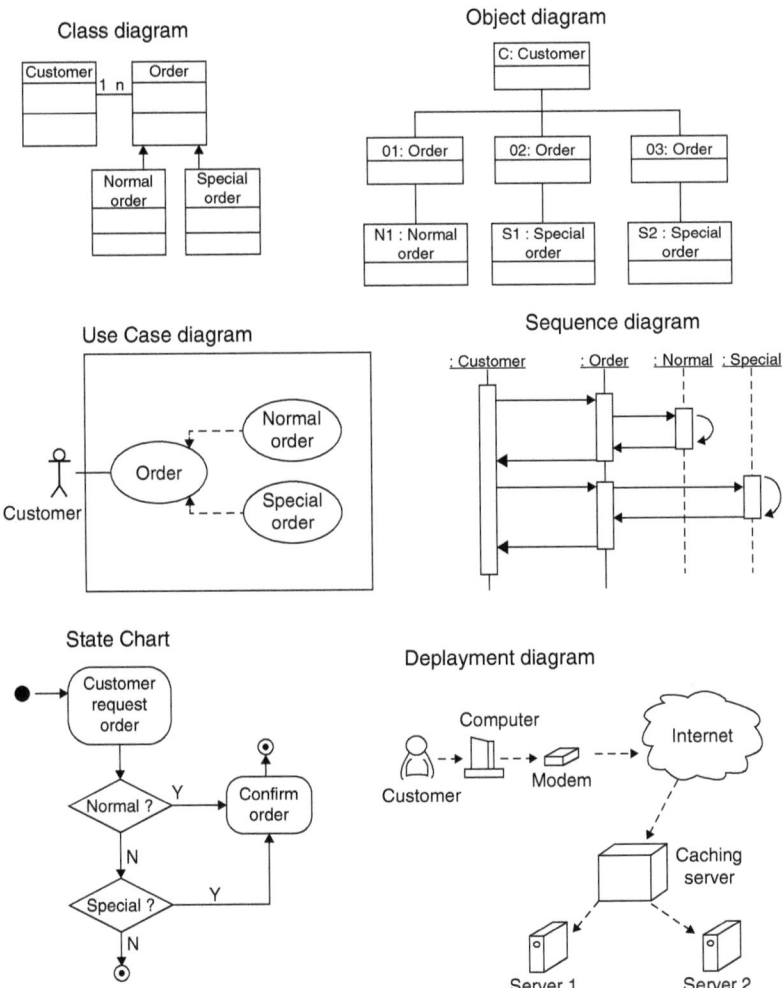

Fig. 2.5 Some examples of UML diagrams

2.5 StateChart

StateChart is an advanced version of traditional State Machine that incorporates extended features. StateChart has notations for Initial (Default) and Final (Termination) states (Fig. 2.6). Initial state defines (the default) start of the process, while the final state shows the end of a process. The transitions between states are shown with solid lines with arrowheads. Above the transition, message [condition] and action are

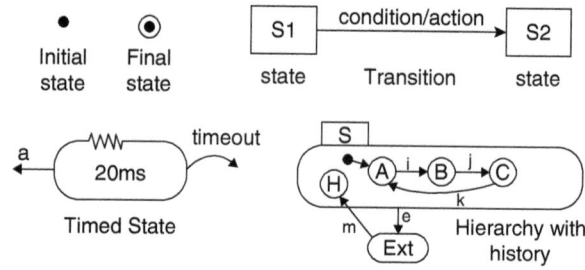

Fig. 2.6 Some notations of StateChart

listed, which are separated by "/". An automata represents a setup with an internal state with input(s) and output(s) where the next state is computed by a function, while the output is computed by another function. Moore or Mealy automata are represented with finite state machines (FSM). A timed automata also contains a model of time. A time variable models the logical clock in the system and is initialized to zero when the system is started. It increases synchronously with the clock rate and expires with a timeout function. Clock constraints, such as guards on edges, are used to restrict the behavior of the automation. A transition represented by and edge can be taken when the clock values satisfy the guard labeled on the edge. Additional invariants make sure that the transition is taken. Clocks may be reset to zero when a transition is taken. Since time needs to be modeled in embedded systems, timer blocks can be used. StateChart provides a special state symbol for timer blocks (Fig. 2.6). If the event "a" does not happen while the system is in the timed state, a timeout occurs, and transition happens toward the timeout transition.

StateChart can also have hierarchy. A group of states in the same level can be enclosed within a package called Super-State. In addition to default and termination mechanism, StateChart also has notation for History that is useful in the hierarchical architecture. Enclosing multiple states within a super-state can hide internal structures from outside world. StateChart also allows Orthogonal Regions that are independent of each other but resides within the same system. An example of orthogonal regions is an oven system that has an independent light with a dedicated switch. Turning the light on and off with its dedicated switch are completely independent of the oven being on or off.

StateChart also allows Fork and Join mechanism. A fork allows a transition to start multiple parallel transitions, while a join allows multiple transitions to merge to a single transition. The join transition must wait until all transitions coming in to join are completed. StateChart also allows concurrency that is a convenient way to describe concurrent events within the same system. In addition, StateChart can also show lifetime of events using variables with logical conditions. Events live until the step following the one in which they are generated. These types of events are called one shot-events. Lifetime of event is a very important concept for resolving conflicts such as Freezing. As embedded systems are typically unsupervised, thus requires self-conflict resolution capability (e.g., unfreeze itself).

In some cases, choices between conflicting transitions are resolved arbitrarily. Thus, StateChart ensures determinate behavior, which is semantics if transition conflicts are resolved and on other sources of undefined behavior exists. However, ES suffers from potential other sources of non-deterministic behavior from the enclosing system (e.g., unpredictable environment).

2.6 Some UML Examples

2.6.1 Desk Lamp UML Diagrams

A simple example of an ES can be a desk lamp that turns on if a push button is pressed and turns off if the button is pressed again. The lamp has two states: On and Off. We can use a Boolean to keep track if the lamp is on or off. A StateChart and a class diagram are shown in Fig. 2.7.

2.6.2 UML Diagrams for an Oven Controller System

An oven system can be decomposed into a unit driving the oven heating coils and control unit for cooking steps. The controller will be responsible for setting the power level and cooking time. The user will set the required settings before Start, then the oven will operate as per settings, then automatically Finish when the cooking time expires. A class diagram and a StateChart diagram are provided in Fig. 2.8.

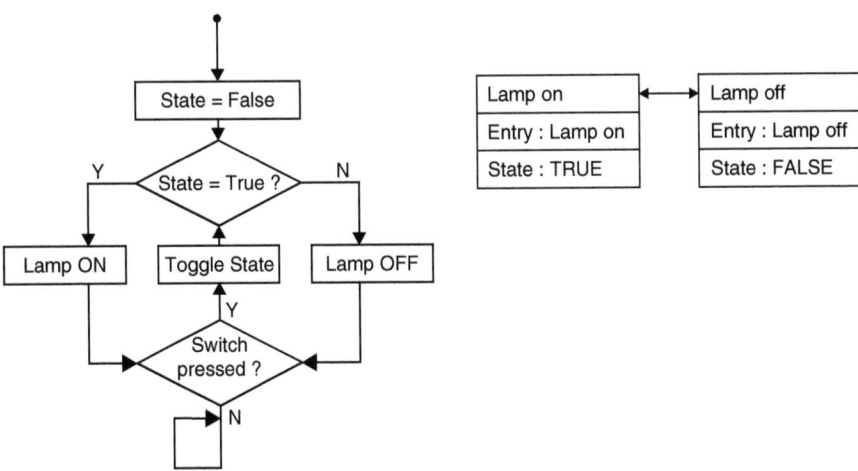

Fig. 2.7 A desk lamp UML diagrams: StateChart and Class diagram

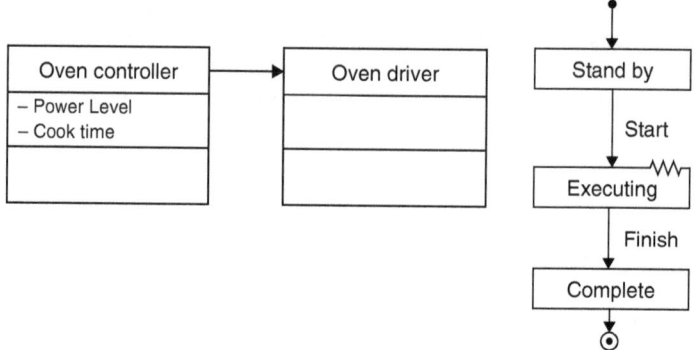

Fig. 2.8 An example oven control system class diagram and StateChart

2.6.3   UML Diagrams for a Car Navigation System

Figure 2.9 shows some UML diagrams for a car navigation system. The two use cases for changing the volume and looking up destination address require different time constraint for real-time operation. The use cases show the details of information that needs to be communicated and required timing constraints for sub-steps.

2.6.4   UML Diagrams for an Audio Message Recording System

A digital audio message recording system allows the user to record audio notes and allows to replay these at a later time. Some related UML diagrams are shown in Fig. 2.10.

2.6.5   UML Diagrams for an Elevator System

Elevator is a complex example of an embedded system. Figure 2.11 shows some UML diagram for an elevator system.

2.7   Design Approaches and Hardware–Software Partitioning

ES requires unified development approach as the design process if more dynamic. Designer must have expertise with both software and hardware to optimize design metrics. In contrast, typical industry jobs require only hardware or software expertise. ES designers must be comfortable with various technologies in order to choose the best for a given application and constraints. The design complexity of ES is increasing, and the time-to-market is reducing. Hence, the traditional design techniques (independent hardware and software design) are challenged. ES utilizes a philosophy called "Hardware–Software Co-design" where the designers consider trade-off in the way hardware and software components of a system work together. The hierarchical layers of hardware and software are together represented as "Hardware–Software Ladder." Recent maturation of synthesis enables this unified view of hardware and software. The choice of hardware versus software for a particular function is simply a tradeoff among various design metrics, like

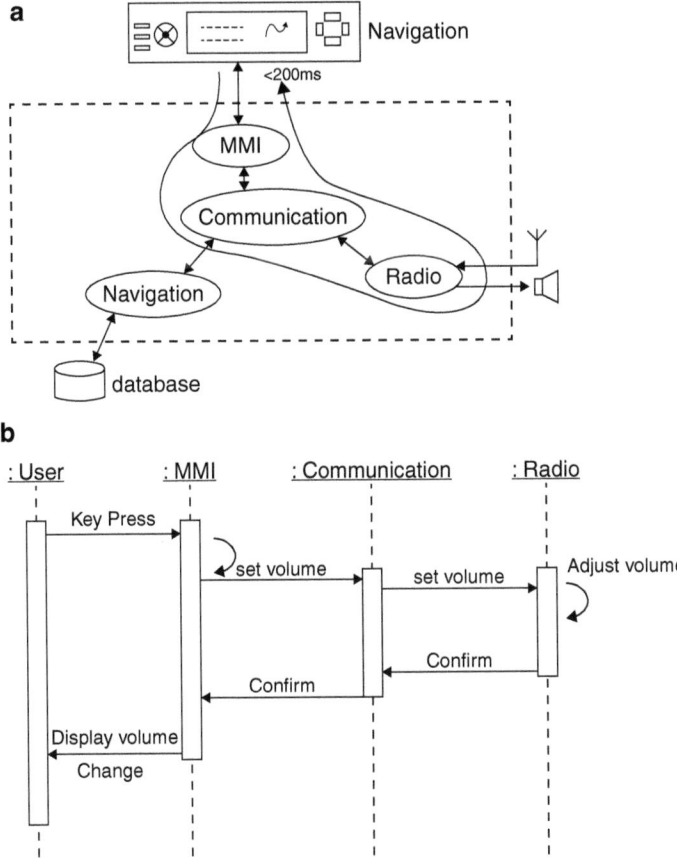

Fig. 2.9 UML diagrams for a car navigation system. (**a**) Use case for change of volume. (**b**) Sequence diagram for change of volume. (**c**) Use case for loop up of destination address. (**d**) Sequence diagram for look up of destination address

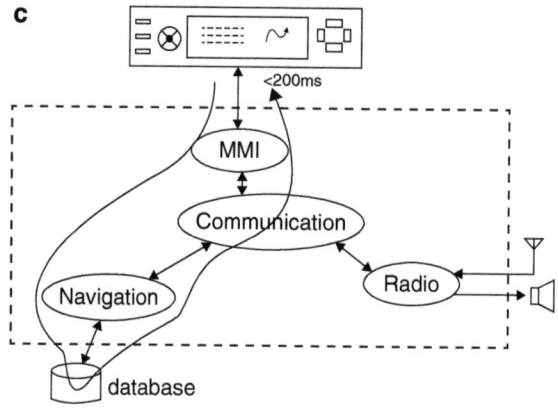

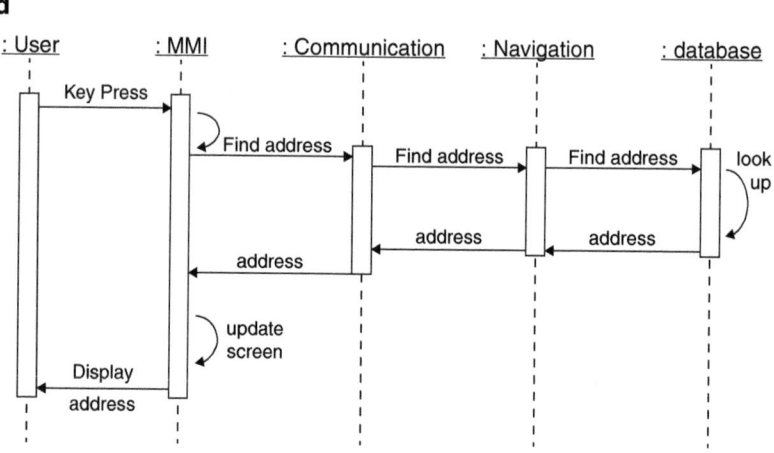

Fig. 2.9 (continued)

performance, power, size, NRE cost, and especially flexibility; there is no fundamental difference between what hardware or software can implement.

ES designers need to be prudent to decide hardware–software partitioning. An
example of this is provided in Fig. 2.12. Here the designer can choose to implement
the band pass filter in hardware domain or in software domain. Both have pros and
cons, and the designer needs to be prudent enough to make a judgment that satisfies
performance while making the product economically viable. Other examples can be
implementing functionality in a software (e.g., microcontroller code) or in a programmable hardware (e.g., FPGA). While the hardware provides advantage of fast
computation, a major disadvantage is the added cost for the hardware.

In recent design trend with the availability of touch sensitive display, a lot of other
ES designers are eliminating hardware switches and buttons for software touch

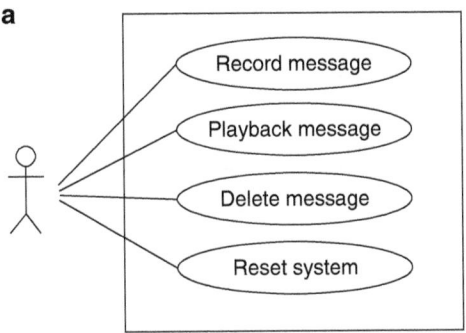

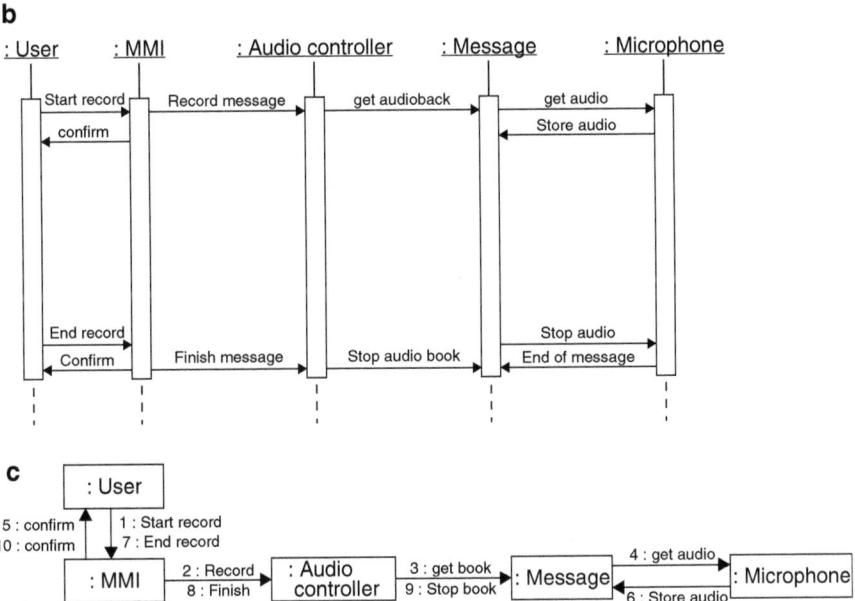

Fig. 2.10 UML diagrams for a voice recording system. (**a**) Use case diagram. (**b**) Sequence diagram for playing messages. (**c**) Collaboration diagram

switches, buttons, and scrolls. For instance, newer model temperature regulators (such as Nest) uses touch sensitive display (software) to interact with the user, such as set the temperature schedule. Also, newer models of cars, such as Tesla Model 3, have eliminated almost all dashboard switches (hardware) and knobs with a touch screen display (software). These types of newer design philosophies require ES designers to think outside of traditions approach of hardware and software boundaries!

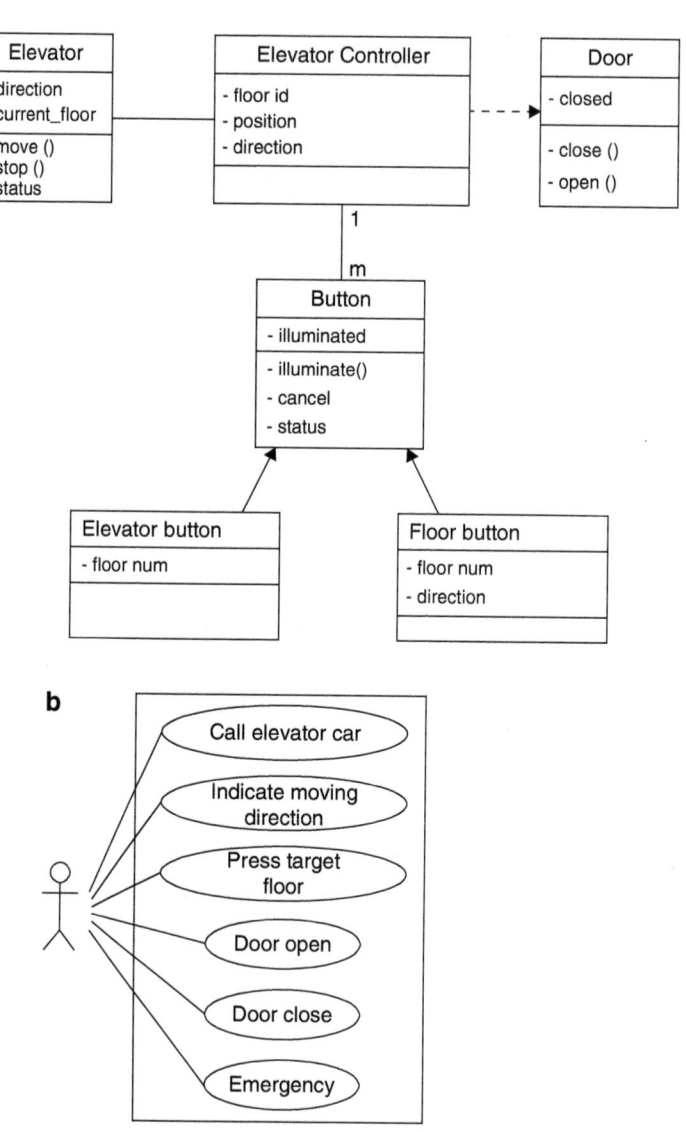

Fig. 2.11 UML diagrams for an elevator system. (**a**) A class diagram. (**b**) A use case diagram. (**c**) A sequence diagram for elevator call. (**d**) A collaboration diagram for elevator call. (**e**) A StateChart for drive control system. (**f**) A StateChart for Door control mechanism

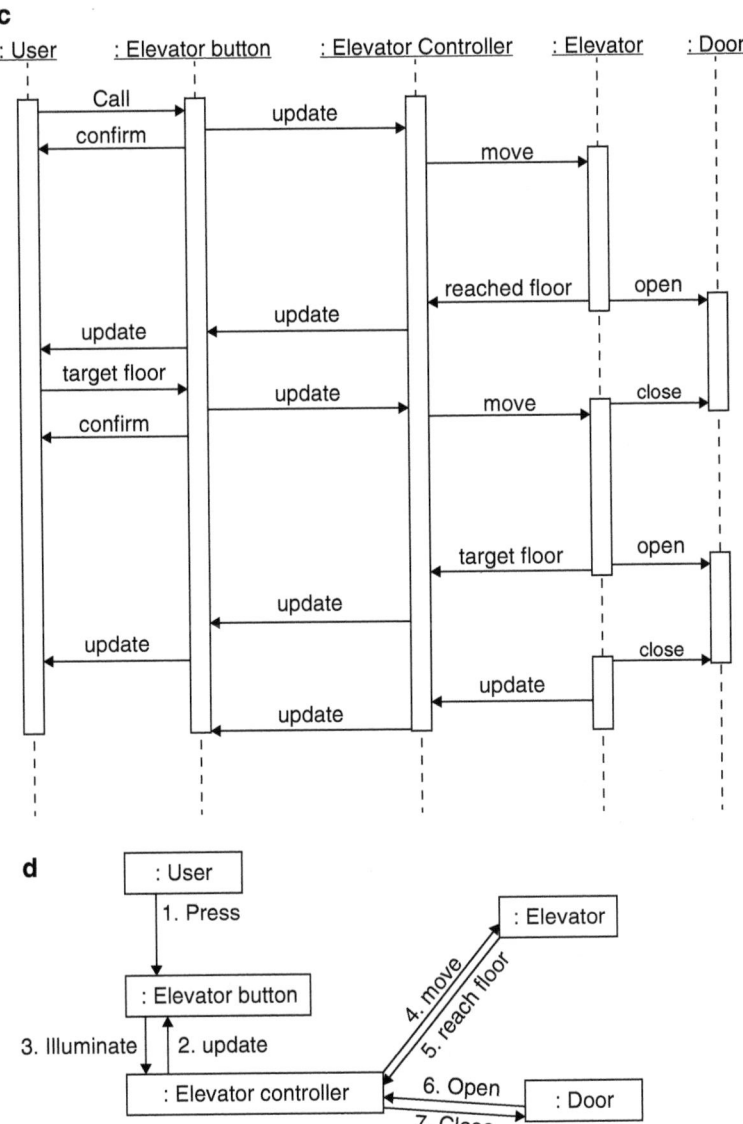

Fig. 2.11 (continued)

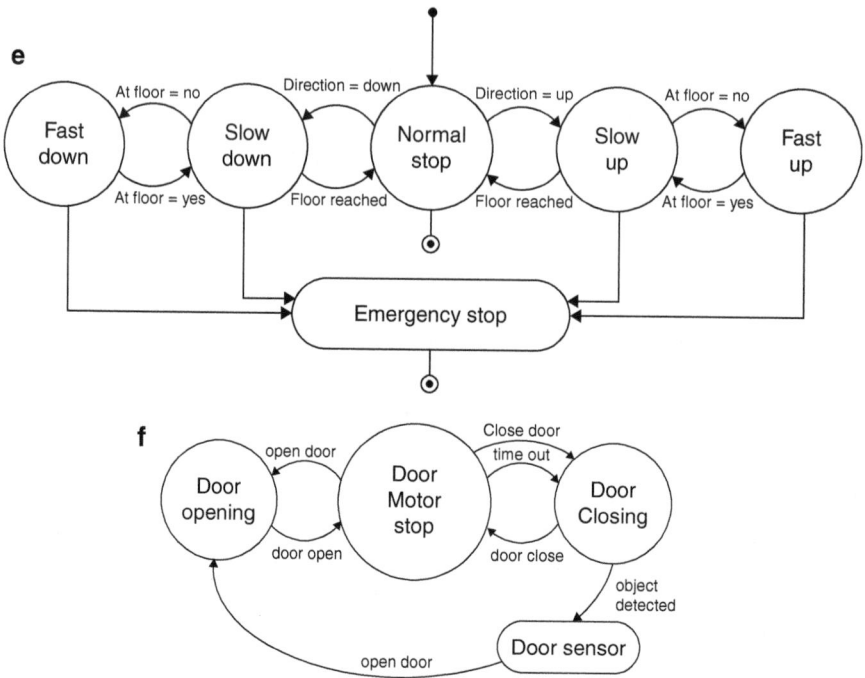

Fig. 2.11 (continued)

2.8   Kopetz ES Design Principles

Kopetz [2] has provided a list of 12 design principles that designers should adhere to as early as system design phase. These design principles are listed below.

1. **Safety considerations:** Safety is an important part of the specification. All systems must consider safety of personnel, equipment, outcomes, and surroundings. This aspect may drive the entire design process. ES in contact with human must be non-hazardous and non-toxic, such as wearables and implants. ES that can impact human life must also be extremely well designed, such as an airplane or a self-driving car.

2. **Precise specifications:** Design hypotheses must be specified precisely and correctly formulated from the beginning. These include expected failures and their probability. Imprecise specification might lead a system to operate outside

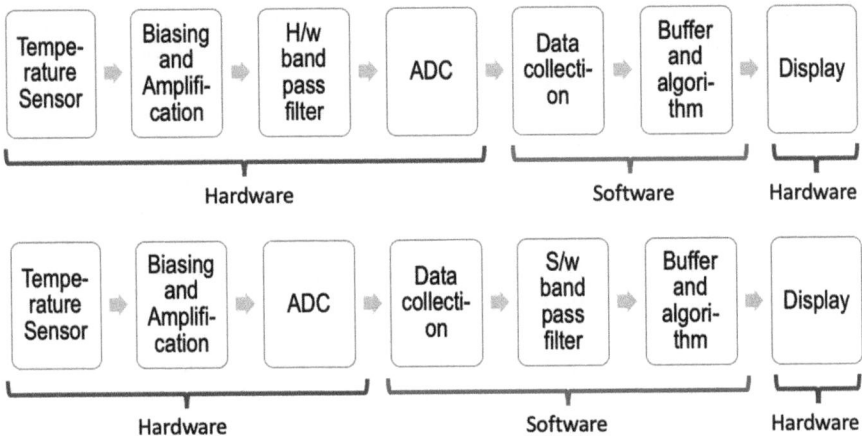

Fig. 2.12 A temperature sensor ES as an example of hardware–software partitioning

of allowed operational boundary. In the case of iterative development, the specifications can have improved precision in subsequent phases.

3. **Fault containment regions (FCRs):** For ES, FCR must be considered from the time of design. Faults in one FCR should not affect other FCRs. For example, when a car crashes on a tree, the front portion of the car should absorb the majority of impact so that the car cabin does not suffer from the impact significantly.

4. **A consistent notion of time and state:** An ES system must establish a consistent notion of time and state. Otherwise, it will be impossible to differentiate between original and follow-up errors.

5. **Well-defined interfaces:** These features improve user experience while hiding the internals of components that are not necessary for the users. Consistent interfaces allow users to quickly grasp new designs from their past experience.

6. **Components fail independently:** It must be ensured that if one or more components fail, they fail independently without leading to failure of other components.

7. **Self-confidence:** Components should consider themselves to be correct, unless two or more other components pretend the contrary to be true. This is the principle of self-confidence.

8. **Fault tolerance:** ES must maintain mechanisms such that they do not create any additional difficulty in explaining the behavior of the system. Fault tolerance mechanisms should be decoupled from the regular function. For instance, if the temperature sensor fails (open or short) in a room heating ES, the system should not be stuck to continuous heating or cooling mode. A simple solution can be if the temperature sensor data is beyond a certain range, the system can flag it as an error condition and turn off both heating and cooling system while raising an alert.

9. **Designed for diagnosis:** The ES must be designed for diagnosis. For example, it has to be possible to identifying existing (but masked) errors. These considerations must be incorporated early in the design phases rather than afterthoughts.

10. **Man-machine interface must be intuitive and forgiving:** Safety should be maintained despite mistakes made by humans by developing man-machine interfaces that are intuitive and forgiving. For instance, a car steering is designed with a small toe in angle of the wheels so that minor movements steering during straight driving does not cause car to change direction. If possible, machines can be optimized to assist humans in case of mistakes, such as the new lane assist technology introduced in some cars.

11. **Keep record:** Every anomaly should be recorded (or logged). These anomalies may be unobservable at the regular interface level. Recording to involve internal effects, otherwise they may be masked by fault-tolerance mechanisms.

12. **Never-give up strategy:** ES should provide a never-give up strategy. ES may have to provide uninterrupted service and becoming unusable with minor issues is unacceptable.

Exercise

Problem 2.1: Draw UML diagrams for a traffic light controller with pedestrian switches.

Problem 2.2: A remote controlled robotic humanoid system composes of two distinct units: a robot unit (hardware/software) that can move freely on a set of wheels, and a remote control unit (e.g., a smart phone app). The two units communicate via a wireless link. The minimum requirements for the robot unit are:

- Basic movements: Move Forward, Move Backward, Turn Left, Turn Right, and Wait.
- Basic human–machine interfacing (HCI): Greet when turned on, Transmit video from a webcam to the remote control unit for real-time monitoring.

Provide the followings in terms of UML:

(a) A list of *Structural Things* that is required at the minimum for the system.
(b) List three *use cases* for the system.
(c) Draw the following UML diagrams:

1. Class diagram
2. Component diagram
3. Object diagram
4. Deployment diagram
5. Use case diagram (at least 3 corresponding to b)
6. Collaboration diagram (at least 3 corresponding to b)
7. StateChart diagram (at least 3 corresponding to b)
8. Sequence diagram (at least 3 corresponding to b)

Problem 2.3: The robotic system is composed of two hardware units: a robot unit that can move on a set of wheels, and a remote control unit. The two units communicate with a dedicated wireless (RF) link. The robot unit must perform the following basic movements at the minimum:

- Move Forward
- Turn Left
- Turn Right
- Wait

In addition, you are free to think of additional functionalities, as you see appropriate for such a system. The objective of this problem is to incorporate as many different UML constructs as possible, while still using them in logical, meaningful ways. In the report, include the followings:

(a) A list of Things (components) and use cases for the system.
(b) The following UML diagrams (use IBM Rational Modeller Rhapsody software, if possible):

  1. Class diagram
  2. Component diagram
  3. Deployment diagram
  4. Case diagram
  5. Interaction diagram
  6. Collaboration diagram
  7. StateChart diagram
  8. Activity diagram

Problem 2.4: You are designing the embedded system of the elevator system at the Engineering Science Building. The building has 3 floors, where each floor can have a maximum of 2 buttons: up and down. Inside the elevator, the user has a choice to select the destination floor, open and close door. The light and fan of the elevator should be automated to turn on only when occupied.

Answer the followings for this problem:

(a) A list of Things (components) and a list of use cases for the system.
(b) Generate these UML diagrams: Class diagram, Component diagram, Deployment diagram, Case diagram, Interaction diagram, Collaboration diagram, StateChart diagram, and Activity diagram.

References

1. Ober, UML-based modelling and verification of real-time systems, Ecole Temps Reel, (2009)
2. H. Kopetz, *Real-Time Systems: Design Principles for Distributed Embedded Applications* (Springer US, 2011)

Chapter 3
Hardware Design

> The technology at the leading-edge changes so rapidly that
> you have to keep current after you get out of school. I think
> probably the most important thing is having good
> fundamentals.—Gordon Moore, co-founder of Intel Inc.

3.1 Microcontroller Vs Microprocessor

Embedded systems overwhelmingly utilize microcontroller units (MCU). Historical microcontrollers were specialized circuits (such as Programmable Logic Device, Programmable ROM) made with array of logic gates. However, newer microcontrollers contain a low-complexity microprocessor core along with some memories and peripherals commonly needed for embedded system designs. Although old generation microcontrollers were relatively simple digital logic gates that can be programmed by the end-users, newer microcontrollers contain very different architecture, namely a microprocessor with its own memory and peripherals on the same chip. Thus, new generation microcontrollers can be called "computer-on-a-chip" and the definition ought to be redefined. In consideration of these recent technologies for MCU, Table 3.1 shows some key differences between a typical microprocessor and a new generation microcontroller.

3.2 History of Microcontrollers

History of microcontroller goes back to first microcontrollers that originated from digital logics. First microcontroller is credited to Texas Instruments (TI) engineers Gary Boone and Michael Cochran who developed TMS1000 in 1971. Early microcontrollers were logic array-based architecture and aimed to integrate multiple digital logics on a single chip. Later Intel developed 4004 and 8048 (microprocessors) that laid the foundation for newer microcontrollers. Instead of a sea of logic arrays, newer microcontroller uses microarchitecture (architecture of a

© Springer Nature Switzerland AG 2021
B. I. Morshed, *Embedded Systems – A Hardware-Software Co-Design Approach*,
https://doi.org/10.1007/978-3-030-66808-2_3

Table 3.1 Key differences between a microprocessor and a microcontroller

Microprocessor	Microcontroller
A microprocessor is the main component of a computer	A microcontroller is itself a small computer
It contains one or more cores with ALU, register set, control unit, etc.	It contains a low-complexity microprocessor core, memory, peripherals, etc.
It has high processing capabilities	It has low processing capabilities
It consumes high power	It consumes low power
It is used for general purpose computing (e.g., personal computer, server)	It is used in embedded systems (e.g., car, modem, TV, IoT, wearables)

microprocessor) that allows persistent software code (firmware) to run on the microcontroller hardware, instead of only hardware-based execution in the logic arrays.

3.3 Various Microcontrollers

In this section, various microcontrollers are introduced with historical perspectives, and their relative advantages and disadvantages.

3.3.1 PLD, PLA, and PAL

The earlier microcontrollers are based on logic arrays or programmable read-only memory (PROM) that contained an array of wire-interconnects that were programmable (burnt for shorts) only once. Based on the program, certain inputs would produce certain outputs. These were known as Programmable Logic Device (PLD), Programmable Logic Array (PLA), or Programmable Array Logic (PAL). The designers would purchase these as components off-the-shelf (COTS) and program layers that pre-existed in the chip. Connections on the chip were either created or destroyed to implement the desired functionality. The advantages of these devices were reduced NRE costs and almost instantiations availability of programmed microcontrollers. But they were power hungry, slow, and had a low processing power.

3.3.2 EPROM and EEPROM

Erasable Programmable Read-Only Memory (EPROM) were the next generation of microcontrollers. It was invented by Dov Frohman of Intel (1971). EPROMs were programmable and erasable under ultra-violet (UV) exposure. The programed chip will retain its data even without power (viz. non-volatile). Typical fabrication process involved floating-gate Field Effect Transistors (FET). It used fused quartz window to allow the UV to reach the transistor gate oxide that would capture charges when exposed to UV, thus enabling the switching property of the FET.

Later, another type of reprogrammable ROM became available that could perform the programming operation with a higher electrical voltage (instead of UV light). These chips were known and Electrically Erasable Programmable Read-Only Memory (EEPROM or E^2PROM). This are also non-volatile memory by implementing floating-gate transistors (FET) that can be programmed with a higher voltage inducing charge in the gates using field emission technique. This technique was invented by Eli Harari at Hughes Aircraft in 1977. Newer non-volatile memory chips still using similar techniques (e.g., Flash Memory). Advanced versions of this techniques are still being researched are Ferroelectric RAM (FeRAM) and Magnetoresistive RAM (MRAM) [1].

3.3.3 FPLA and FPGA

Field Programmable Logic Array (FPLA) was introduced around 1975 by Signetics. It aimed to replace and improve upon multiple logic chips (e.g., AND, OR, NOT, NAND, Multiplexor, etc.) that were typically used at that time to develop embedded digital controllers. FPLA integrated many logic gates in a single chip and allowed the users to be able to program the interconnects among these logic gates based on their needs. An FPLA chip would contain multiple input logic lines that feed to a series of AND (input gates), followed by OR and XOR gates (summing matrix). The interconnects were made with programmable links. Although visionary, the product did not gain much market-share primarily due to marketing challenges. This technology was picked up by US Navy in the 1980s and improved significantly.

Later, in the 1980s, a newer technology, named Field Programmable Gate Array (FPGA) came to market. There were two main manufacturers of FPGA: Altera (founded in 1983) and Xilinx (founded in 1985). These companies advanced FPLA to a new generation of FPGA. These devices integrated basic logic gate combinations in units called Configurable Logic Blocks (CLB). These logic blocks are designed and optimized to provide reconfigurable logic gates. Programmable logic blocks were invented by David W. Page and LuVerne R. Peterson, and patented in 1985 [2, 3].

The CLBs typically consist of some Look-up tables (LUTs), followed by combinational logic blocks like Full Adders, and sequential logic blocks like D-Flip-Flop. The CLBs are connected with programmable routing wires (i.e., switch box). Most FPGAs also have memories (e.g., RAM block), Delay Locked Loop (DLL), times, and input/output (I/O) cells. The users can design custom logics with Hardware Description Language (HDL) code that can be synthesized to be downloaded in the FPGA to perform control algorithms.

The traditional FPGAs would only allow hardware-based microcontroller designs. But newer FPGAs contain built-in microprocessors in addition to reconfigurable hardware. For instance, newer Altera FPGAs (e.g., Cyclone series) contain IP Cores (viz. Soft Microcontroller) named Nios II. This expands the capabilities of FPGAs to perform signal processing with both hardware and software as needed. Due to fast speed (of hardware), FPGAs are suitable for high-speed processing and Finite State Machines (FSM). But they are typically expensive compared to other simpler microcontrollers and complex, hence the design requirement must warrant use of FPGA for an embedded system application.

3.3.4 ASIC

Microcontrollers could be designed with custom IC fabrication process such as Application Specific Integrated Circuit (ASIC). An ASIC is a full-custom IC device such as a microcontroller that is designed for specific applications. With ASIC design, the engineers have the complete control over the mask layers used to fabricate the silicon chip, which means the designer can only include logic gates that are needed. Furthermore, they can also handcraft the dimensions of the individual transistors to minimize power consumption and achieve high level of functionalities with the fewest possible transistors. This leads to the most optimal microcontrollers both in terms of energy consumption and performance. A full-custom IC results in the minimal waste of silicon chip area. However, designing ASIC chip is a highly complex, time consuming, and expensive process. Thus, this approach leads to a huge NRE, thus not suitable for most embedded system applications. Only exceptions are when the volume of production is large enough to compensate for the NRE, or the application requires high optimality while the expense is irrelevant. Examples may include implantable pacemakers, brain implants, and specific-purpose and delicate bioelectric signal acquisition chip.

Full-custom ASIC allows all layers of chip design to be optimized by the designers, which includes placement of transistors, sizing of transistors, and routing of interconnect wires. This leads to even optimize performance, smallest size, and lowest power consumption. However, disadvantages such as high NRE cost (often more than $500 k) and long time-to-market often offsets the advantages. Semi-custom ASIC are faster to design and develop, hence leads to lower NRE. In this case, lower layers of chip design are fully or partially built. Chip designers only customize the routing wires and maybe place some additional blocks as needed. This technique is also known as Application Specific Standard Parts (ASSP). The advantages of this technique are good performance gain of the chip, small footprint, compact IC, and relatively lower NRE cost than a full-custom implementation (e.g., between $10 k and 100 k). The disadvantages include still relatively larger time-to-market and NRE (compared to FPGA or COTS microcontrollers).

3.3.5 Microcontroller Unit (MCU or uC)

Newer microcontroller units (MCU) are a "computer on a chip"! This is the most popular types of microcontrollers for complex embedded system design. Most commonly known microcontrollers are this type. The MCU contains a microprocessor unit, one or more oscillators (clock, timer), memory (non-volatile program memory and volatile RAM), Analog to digital (A/D) converter, and input/output (I/O) ports. Some MCU also includes some Digital Signal Processing (DSP) hardware such as digital filters, Fast Fourier Transform (FFT), and digital to analog (D/A) converter. MCUs have distinct advantages of being very inexpensive and allowing a very short development time (i.e., low NRE and time-to-market). The disadvantages are mainly related low computational resources, lower speed, and lower performance.

3.3.5.1 ATMEL

Atmel MCU, developed by Atmel Corporation, are excellent for low-power and low-resource embedded systems. A popular Atmel MCU model is AVR which provides products (e.g., ATtiny, ATmega) in 8, 16, and 32-bit versions. The IP core is a derivative of Intel 8051 microprocessor architecture. These MCUs supports up to 20 MHz clock rate. Memory range are between 0.5 and 16 kB of Flash. The MCUs typically include I/O ports like Pulse Width Modulation (PWM), Analog to Digital Converter (ADC), Universal Synchronous/Asynchronous Receiver/Transmitter (USART), and Serial Peripheral Interface (SPI). These MCUs can be very small-in-size (as low as 8-pin packages) and consumes very low power - which are ideal for a lot of ES applications. However, these are low-complexity MCU, hence cannot handle high workload or processing requirements. They are also very inexpensive. Atmel provides a free Integrated Development Environment (IDE) software, called "Atmel Studio," for programming and debugging of these MCUs. Recently, Atmel has been acquired by Microchip Technology.

3.3.5.2 PIC

A different type of microcontrollers known as Programmable Interface Controller (PIC), developed by Microchip Technology, is an efficient low-power, low-resource, and low-cost option. Newer PIC MCU contains 16-bit CPU in Harvard architecture with RISC ISA. Early models used ROM or EPROM for memory, but newer models contain flash memories. Most of these MCU have common I/O ports such as I2C, SPI, USB, ADC, PWM, and UART. The IDE for this MCU is MPLab.

3.3.5.3 MSP

Texas Instruments (TI) had developed an extremely low-power microcontroller series (MSP 430) that is an attractive option for very low power operation with this highly capable mixed-signal microcontroller. It uses Von Neumann Architecture with 16-bit RISC (MIPS-like) ISA. It has 16 registers and uses Little Endian memory. The rich configuration of these series includes a range of flash memory from 1 and 512 kB, ROM options from 1 and 66 kB, RAM options of 128–10 kB, and ADC options of 10, 12, or 14 bit SAR type, as well as 16 or 24 bit Sigma–Delta type. Typical I/O ports include UART, I2C, USB, and even capacitive touch sensing port. The IDE tool for this series is called Code Composer Studio.

3.3.5.4 ARM

ARM (Acorn RISC Machine) is licensed by ARM Holdings, and manufactured by various chip companies including Atmel, Silicon labs, TI, Kinetis, Freescale, Cypress, STMicroelectronics, Samsung, Renesas Electronics Corporation (which is a spin-off of NEC), etc. There are various ARM core designs that ranges from ARMv1 to ARMv8. They are either 16, 32, or 64-bit processors (i.e., widths of the registers). As the name suggest, the processor uses ARM ISA. There are both flavors of architecture available, for example ARMv7 is Von Neumann, whereas ARMv9 is Harvard architecture. The chips come in three typical configurations: Application (A), Real-time (R), and Microcontroller (M). Almost 95% of smartphones are using ARM processors. The IDE for ARM processors is called ARM Development Studio.

3.3.5.5 ATOM

Intel Corp. has suffered from its legacy issues (such as backward compatibility, CISC ISA) that complicated its entrance to microcontroller domain. Recently they have developed an ultra-low power MCU with 45 and 22 nm CMOS technologies that has a high processing power. The processor is named as ATOM. This supports 32 and 64-bit hardware, and compatible with ARMv7 or higher MCUs. I/O ports available are SPI, JTAG, SMBus, PCI, etc. They also contain some advanced peripherals such as RTC, WDT, and DDR2 memory. The IDE for this MCU is Eclipse.

3.3.5.6 PSoC

Programmable System on a Chip (PSoC) has been developed by Cypress Semiconductor. These processors are low power and contains reconfigurable analog units that leads to a unique design potential reducing the component counts on the PCB and simplifying the design process. Nonetheless the processors consume higher power compared to competitors (such as MSP 430) due to analog components. Typical current consumption in active mode is ~2 mA, while sleep mode can reduce current to as low as 1–3 μA. There are four models of PSoC: PSoC1 (contains a 8-bit M8C core), PSoC3 (contains a 8-bit 8051 core), PSoC4 (contains a 32-bit ARM Cortex M0 core at 48 MHz), and PSoC5 (contains a 32-bit ARM Cortex M3 that can run at speeds up to 80 MHz). They typically contain ADC, DAC, Analog op-amp, I2C, SPI, USB, UART, and CAN ports. The IDE needed for this PSoC1 is called PSoC Designer, while for the rest of the PSoC, the IDE needed is PSoC Creator.

3.3.6 Popular Commercial Prototype Boards

There are a variety of prototype boards that are readily available in market that utilizes these MCUs. The most common of these is Arduino. This section briefly discusses a few of the prototype boards.

3.3.6.1 Arduino

Arduino was developed with open source hardware/software concept that has led to its high popularity and low price. It is considered an entry level MCU prototyping board. Popular models are Arduino Uno, Arduino nano, and Arduino mega. They use AVR microcontrollers such as ATmega8, ATmega168, etc. The clock speeds are typically 8 or 16 MHz. The board contains a boot-loader and on-board flash memory of the MCU that makes is simple and easy to use. Also, there are many shield boards (e.g., GPS, Ethernet, LCD, Bluetooth, WiFi) available for most Arduinos that makes rapid prototyping possible. The I/O ports are Analog, GPIO, USB, I2C, SPI, RS-232, PWM, etc. The IDE is called Arduino Sketch.

3.3.6.2 Raspberry Pi

Raspberry Pi Foundation has developed these boards for promoting educational development and licenses the technology. Common models are: Raspberry Pi 0 (contains ARM MCU with 1 GHz clock and 512 MB RAM), Raspberry Pi 1 A (contains ARMv6 core with 700 MHz clock and 256 or 512 MB RAM), and Raspberry Pi 1 B (contains ARMv7 core with 900 MHz clock and 1 GB RAM). They contain VideoCore IV GPU and USB and audio/video ports. The PIs are programmable with Python and supports C/C++, Java, Perl, and Ruby.

3.3.7   Selection of MCU

It is not straightforward to determine the best suited MCU for an application. Many factors must be carefully considered that includes MCU speed, memory size, power consumption, voltage requirement, required ports, etc. In some high-speed applications, FPGA or a combination of FPGA with SoC might be suitable. For simple cases, Arduino is sufficient and is the most popular among beginners. Thus, we will constrain our discussion within this board, but will provide pointers when Arduino is not sufficient, and alternatives must be assessed.

3.3.8   Programing of MCU

Some MCU prototyping boards, such as Arduino, a simple connection to computer with a USB cable is sufficient. However, some microcontrollers require dedicated programming hardware, such as PICkit 2 for PIC MCU or MiniProg3 for PSoC MCU. One advantage of these programming hardware is that it can allow the programmer to debug the content of the MCU registers and variables (called in-circuit debugger), which can be very effective to debug complex code. Some boards can contain these hardware within the prototyping board (such as Renesas Prototyping boards). They typically use ICMS or JTAG technique for programming and debugging. The IDE supplied by the vendor can allow access to the in-circuit debugger features.

3.4   Analog Front End (AFE)

We live in an analog world! To interact with this analog world, we need analog sensors. Although some sensors appear to be digital, in all probability the sensor contains a machoism to convert sensed analog signals to digital output. Almost all analog sensor requires amplification of the signal, noise suppression, and filtering before it can be digitized using Analog to Digital Converter (ADC). This also requires the sampling to follow Nyquist criteria; however, oversampling is more practical in most cases. For this purpose, it is important that ES designers are capable of designing Analog Front End (AFE). AFE stages can include biasing, amplification, filtering, and analog multiplexor (AMux). On the other hand, for digital sensors (sensors with digital data output) only requires proper data communication port at the MCU, as well as the corresponding digital communication protocol (e.g., communication port driver library) in the MCU's firmware.

Real world interfacing for signals, such as temperature, humidity, pressure, voltages, and currents, which are analog, requires AFE design. Thus, if we want our embedded applications to interact with the outside world, we need to learn to

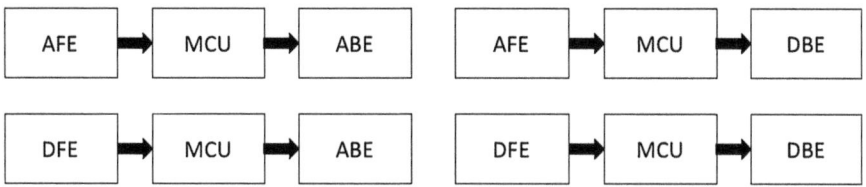

Fig. 3.1 Various configurations of front end and back end for ES

Fig. 3.2 An example
schematic of a temperature
sensor (AFE) connected to
an MCU with an LCD
(DBE)

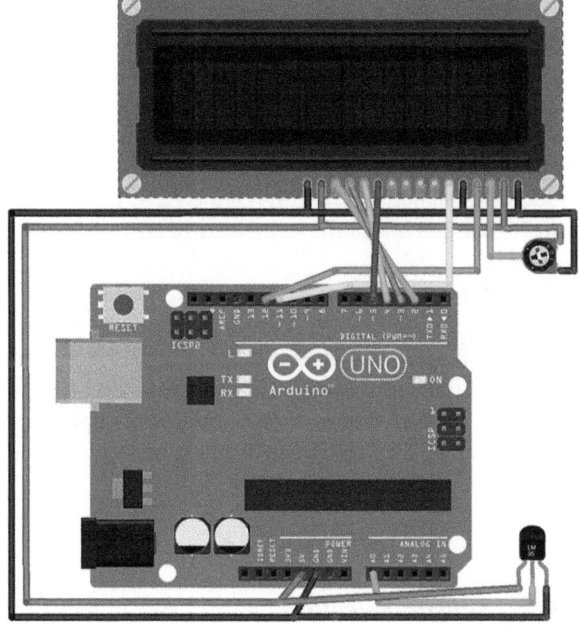

process and interpret analog information and convert it to digital data so that the
MCU can process it and possibly produce an output again through analog or digital
actuators. While ADC is used to convert the analog input signal to digital data, a
Digital to Analog Converter (DAC) modulates at the back end the digital data to
analog signals for the "real" world. The possible arrangements of these analog
(A) and Digital (D) front end (FE) and back end (BE) can be arranged as shown in
Fig. 3.1.

An example of schematic is shown in Fig. 3.2, where an analog temperature
sensor (serving as the AFE) is connected to an Arduino (MCU) and a digital LCD
(serving as the DBE). The MCU will contain the software (firmware) to sample the
temperature signal (from A0 port), convert the voltage data to the correct digital
value (through software codes), and display this data by transmitting it to the LCD
(using digital communication protocol). In this example, the AFE only contains the

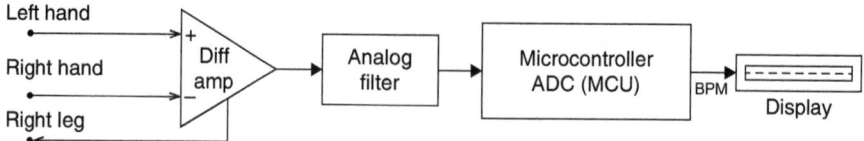

Fig. 3.3 Another example of a heart rate display device from ECG signals with the AFE constituting of electrodes, an amplifier, and an analog filter connected to an MCU and an LCD as the DBE

temperature sensor, although later we will see that analog signal processing is typically required to improve the system performance.

Another example of ES shown as a block diagram in Fig. 3.3 for heart rate capture system from electrocardiogram (ECG or EKG) signals. In this example, the AFE contains ECG electrodes, an amplifier (Amp), and an analog filter. The MCU (6812) collects this analog signal through the ADC port, computes the heart rate from this ECG signal and sends the computed heart rate to the LCD (constituting DBE) to display it.

As mentioned before, sensors are required to sense physical parameters. In fact, sensors contain the relevant transducers that can convert the physical information to electronic signals. Any physical stimuli must be converted to electrical signals (i.e., voltage, current, or change of impedance). Some examples of analog sensors are thermistor, pressure sensors, flex-sensor, microphone, and biopotential electrodes. Some examples of digital sensors are camera, ultrasound sensor, proximity sensor, and global positioning system (GPS).

3.4.1   Sensor Selection

The first step of AFE design is to select a proper sensor. As this process widely varies based on the sensor type and application, we will use an example of a temperature data acquisition system (DAS). For this exercise, let's say we selected a temperature sensor that produces resistance change. This type of sensors will produce a change of resistance with change of temperature. For selection of this sensor, one important consideration is the transfer curve. An example is shown in Fig. 3.4.

In this case, the sensor has a negative temperature coefficient (NTC) as the resistance decreases for increased temperature. The curve also shows that the sensor behaves non-linearly. Some datasheet will include the transfer curve, which needs to be considered during the software coding by compensating the non-linearity of the curve. In some cases, piecewise linear approximation can be used for simplification, but the range or accuracy needs to be sacrificed. Furthermore, the datasheet needs to be carefully considered for other important specifications such as operating voltage and currents, power consumption, operating temperature range, sensitivity of resistance, noise characteristics (such as internal flicker noise), hysteresis, size,

Fig. 3.4 The transfer curve of a negative temperature coefficient (NTC) sensor

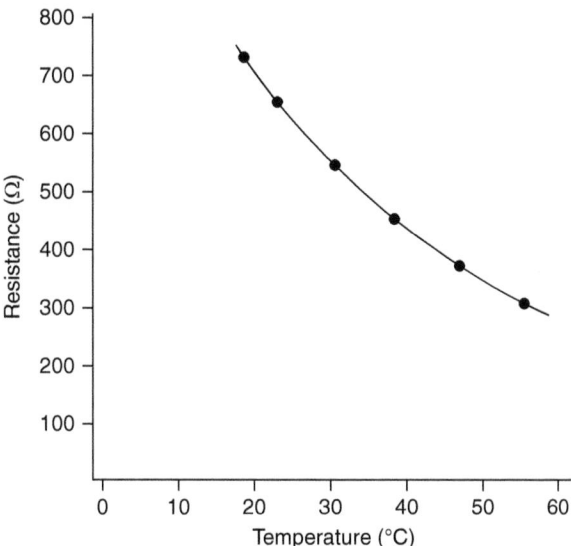

packaging, soldering requirements, and sensor specific parameters (e.g., directivity, transmission distance).

During the sensor selection process, datasheets of the candidate sensors must be carefully considered. Hence, ES hardware designer must be cognizant of the important parameters to compare for the target design. For example, comparing the transfer curve of similar temperature sensors might allow narrowing down the selection process based on the range of temperature to be sensed and the output voltages produced.

To assist in the selection process, a table can be prepared with important parameters to be compared. An example table is shown in Table 3.2, based on arbitrary data. For a room temperature sensor, which type of temperature sensor will be suitable?

3.4.2 AFE Design

AFE design can be simple or complex depending on the sensor type and application. Although using a DFE is simpler for the ES designer, but for optimal performance with lowest unit cost and miniaturization of the complete hardware, it might be advantageous to develop in-house AFE. The example in 3B had a very simple AFE, while the example in 3B is relatively complicated. As an example, here we will consider designing a temperature sensor AFE for room heating system, where the example 3B is not sufficient. Firstly, to minimize cost and size, we us assume that we selected a thermistor for our design. The thermistor requires a biasing for proper

Table 3.2 A comparison of various types of temperature sensors

Attribute	Thermocouple	RTD	Thermistor
Cost	Low	High	Low
Temp. Range	−350 to +3200 °F	−400 to +1200 °F	−100 to +500 °F
Stability	Good	Excellent	Poor
Accuracy	Medium	High	Medium
Precision	Poor	Excellent	Good
Sensitivity	Low	Medium	High
Latency	Fast	Medium	Fast
Linearity	Fair	Good	Poor
Self heating	No	Low	High
Package size	Medium	Small	Small

Fig. 3.5 A simple biasing scheme for the temperature sensor

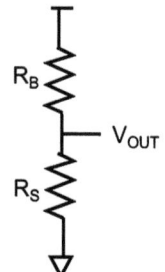

operation. The biasing scheme can be a simple biasing (Fig. 3.5), a Wheatstone Bridge circuit biasing (Fig. 3.6), or other schemes (datasheet should be consulted).

However, the outputs of these biased circuits are typically not sufficient to feed to ADC. Let us investigate the simple biasing first. The output voltage of this scheme can be calculated using the voltage divider equation:

$$V_{OUT} = \frac{R_S}{R_B + R_S} V_{DD.}$$

Assuming the change of R_S is ΔR for ΔT temperature change from an initial temperature T_1 to a final temperature T_2, the change of output voltage can be written as:

$$\Delta V_{OUT} = \frac{R_B \Delta R}{(R_B + R_{S1})(R_B + R_{S1} + \Delta R)} V_{DD}$$

It is evident from this equation that the change of output voltage will be small as typically the factor $R_B \Delta R / ((R_B + R_{S1})(R_B + R_{S1} + \Delta R)) << 1$. Thus, this output cannot be directly fed to an ADC, as the sensitivity will be unacceptably poor. We can rewrite this in terms of sensitivity $|\Delta V / \Delta R_S|$ as follows:

Fig. 3.6 Wheatstone
Bridge circuit configuration
for the temperature sensor

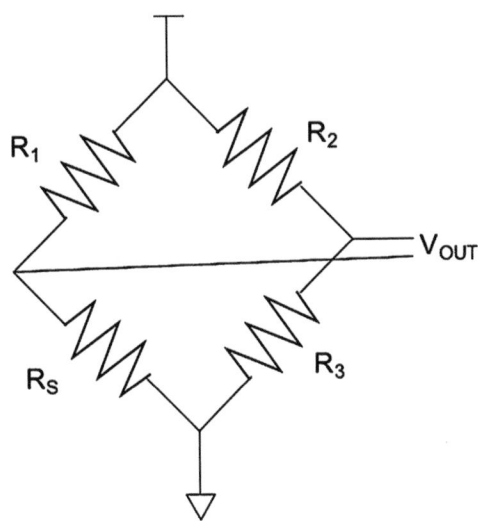

$$\left|\frac{\Delta V}{\Delta R_S}\right| = \frac{R_B V_{DD}}{(R_B + R_{S1})(R_B + R_{S1} + \Delta R)}$$

If the $\Delta R < <(R_B + R_{S1})$, we can simplify this as follows:

$$\left|\frac{\Delta V}{\Delta R_S}\right| = \frac{R_B V_{DD}}{(R_B + R_{S1})^2}$$

This expression suggests that sensitivity can be improved by increasing V_{DD}, and sensitivity will be maximum for when R_B is close to R_{S1} (or R_{S2}). Nonetheless, the sensitivity of a typical simple bias scheme is not sufficient to connect the output directly to the ADC input, rather an amplification stage is often required.

On the other hand, for the Wheatstone Bridge configuration, the output voltage can be written as:

$$\frac{V_{OUT}}{V_{DD}} = \frac{R_2 R_S - R_1 R_3}{(R_1 + R_S)(R_2 + R_3)}$$

For the same ΔR and ΔT, the output voltage difference can be written as:

$$\frac{\Delta V_{OUT}}{V_{OUT1}} = \frac{R_1(R_1 R_{S2} - R_2 R_{S1}) - R_1 R_3(R_{S1} - R_{S2}) + R_{S1} R_{S2}(R_1 - R_2)}{(R_1 R_{S1} - R_1 R_3)(R_1 + R_{S2})}$$

In a typical design, the same values are chosen for R_1 and R_2. In that case, the equation simplifies to:

$$\frac{\Delta V_{OUT}}{V_{OUT1}} = \frac{(R_1 - R_3)R_1 \Delta R_S}{(R_1 R_{S1} - R_1 R_3)(R_1 + R_{S1} - \Delta R_S)}$$

This shows that the change of ΔV_{OUT} can be large when R_{S1} value is close to R_3. Thus, this biasing is sometimes preferred over simple biasing. As a drawback of this scheme, we will need a differential amplifier to amplify the output of the Wheatstone Bridge as the output cannot be directly fed to the ADC. However, if the design already uses op-amp (e.g., for noise filtering), then we can use op-amp for amplification of signal from simple biased circuit which allows more control of analog signal processing in the AFE. For this, we can choose op-amp packings with 2 or 4 op-amps with marginal increase in cost.

To explore this design (simple biasing with an op-amp amplifier), let us assume the temperature sensor resistance is $R_{S1} = 10$ kΩ at $T_1 = 60$ °F initially, which changes to $R_{S2} = 9$ kΩ at $T_2 = 90$ °F. If the supply voltage is $V_{DD} = 3$ V (unipolar), let us compute how this simple bias will work for an 8-bit ADC.

We need to be aware of the fact that op-amp design will require a three voltage levels: positive (V_{DD}), negative (V_{SS}), and mid-rail (V_{MID}). This can be achieved by using a bipolar supply (serving the V_{DD} and V_{SS}) and ground (serving as V_{MID}). For unipolar power supply (as in this example), thus we need to generate a mid-rail voltage (i.e., 1.5 V) and use that as the V_{MID} terminal (aka V_{REF}) of the circuit. We can generate this V_{MID} by simply using a voltage divider with equal value of the resistors. However, this simple approach has a drawback as this will drain current continuously while the device is powered. For low power ES such as battery powered system, this constant power drain needs to be minimized. We can use a very high value resistors to reduce this power wastage. But a high value of resistor will result in a very low current flow path which might not be able to source or sink enough current for the regular operation of current. To understand the problem, let us pick 1 kΩ resistors for this voltage divider. Thus, as shown in the left side of Fig. 3.7, the total resistance is 2 kΩ and constant current through this is 1.5 mA and consuming only 4.5 mW of power. But consider that during the normal operation

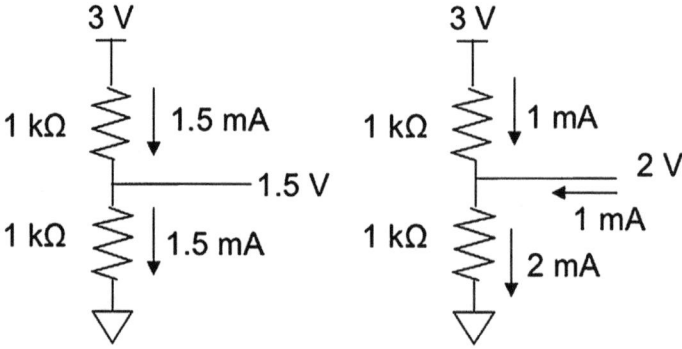

Fig. 3.7 Voltage divider to generate mid-rail voltage (V_{MID}), and an example of voltage swing

Fig. 3.8 Voltage divider
followed by a voltage
follower op-amp to generate
V_{MID}

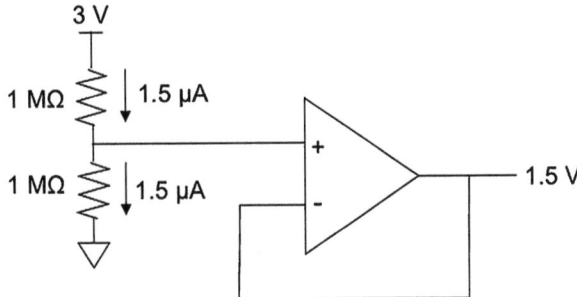

of the circuit, the current needs to be sourced or sinked through this terminal. Let's say the current of 1 mA needs to be sinked through connected terminal of this mid-rail voltage (V_{MID}). In that case, the 1 kΩ resistor connected to the mid-rail and V_{SS} will pass more current than the resistor connected to V_{DD}! For this condition, the V_{MID} will swing up to 2 V (as shown on the right side of Fig. 3.7), resulting 1 mA current through the top resistor, while 2 mA current through the bottom resistor. This means, V_{MID} will swing about 33%, which is unacceptable!

To resolve this issue, most designs will use an op-amp voltage follower setup as shown in Fig. 3.8. In this design, only 1.5 µA current is consumed by the resistive branch (op-amp current consumption is negligible), whereas the output of op-amp can safely source or sink mA level currents without compromising the output voltage. As this requires one more op-amp, quad op-amp packing is a very attractive option for AFE.

Now, let us go back to our temperature sensor biasing circuit. To select a biasing resistor value, we need to remember that a low biasing resistance will lead to a higher current through the temperature sensor. Not only that will increase power consumption, it can also produce localized heat, so that the temperature sensor reading of the ambient will be compromised. On the other hand, choosing a very high biasing resistor might lead to insufficient current through the temperature sensor for proper operation. Furthermore, in general, lower current will lead to higher noise in output. Let us say, after some consideration and consulting with the datasheet, we choose the bias resistance to be $R_B = 10$ kΩ (the same value of R_{S1}).

Now, using the equation mentioned before, we can find out that the output voltage at 60 °F will be 1.5 V and at 90 °F will be 1.42 V. Thus, the output voltage will swing 0.08 V for these two temperatures. As the voltage resolution of an 8-bit ADC for 3 V supply is 11.7 mV, the total number of quantization steps for this voltage swing (i.e., for 30 °F temperature swing) will be $0.08/0.0117 = 6.8$, i.e., 7 steps. So, each quantization step will represent $30/7 = 4.3$ °F, which is not acceptable. In other words, the temperature can be controlled in steps of 4 °F!

To improve the design, we can use a higher bit ADC, but that will increase cost and noise will become an issue for higher bit ADC. The other easier route is to amplify the output of simple bias using an op-amp. This is also preferred due to robustness against noise. The objective here is the amplify the swing of 0.08 V

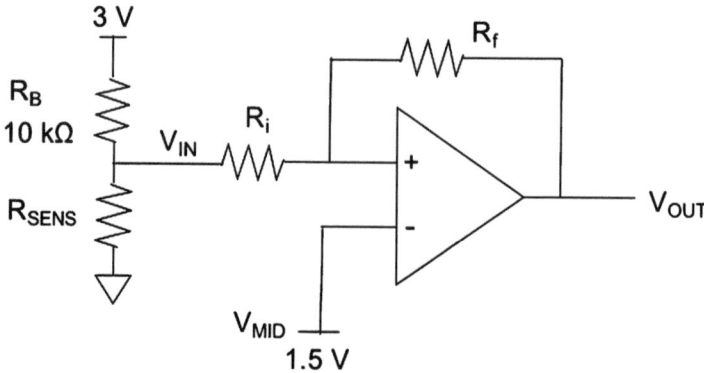

Fig. 3.9 Inverting op-amp to amplify the voltage swing of the simple bias

toward 3 V (full range of supply) without clipping the output voltage. Figure 3.9 shows an inverting amplifier configuration to amplify the output. It is a common choice due to simplicity of the circuit and stable operation. We can also choose to use a non-inverting amplifier instead.

Let us compute the values of R_i and R_f. For the computation of op-amp, we must refer all voltages with respect the mid-rail voltage (1.5 V in our case). Thus, V_{DD} and V_{SS} is +1.5 V and −1.5 V, whereas V_{MID} is 0 V. The variation of the input voltage $V_{IN2}-V_{IN1} = 1.42-1.5$ V $= -0.8$ V. Our intended output voltage swing is 0–1.5 V (the inverting amplifier will invert the input). Using the equation for inverting amplifier gain for the Fig. 3.9:

$$A_v = -\frac{R_f}{R_i}$$

$$=> \frac{1.5}{0.08} = \frac{R_f}{R_i}$$

$$=> R_f = 18.75R_i$$

Now, we can choose a suitable value of R_i. Let us use 22 kΩ for this resistor, then the value for R_f will be 412.5 kΩ. As this resistance value is not available in E-12 series (which is standard values for 10% tolerance resistances namely 1.0, 1.2, 1.5, 1.8, 2.2, 2.7, 3.3, 3.9, 4.7, 5.6, 6.8, and 8.2). Thus, we can select 390 kΩ as R_f. Note that we should select a lower value than the calculated value so that the output does not saturate (or clip) at the maximum range. Also note that for R_B, it is rather preferable to use 1% or 5% tolerance resistance, as the output will be sensitive to its actual value.

The gain for 390 kΩ will be -17.7, which is acceptable. With this value, the V_{OUT} will swing from 0 to 1.416 V (with respect to V_{MID}). The output swing of 1.416 V means $1.416/0.0117 = 121$ steps of 8-bit ADC. Thus, each quantitation step will represent $30/121 = 0.25\ °F$, which is reasonable for most room heating system!

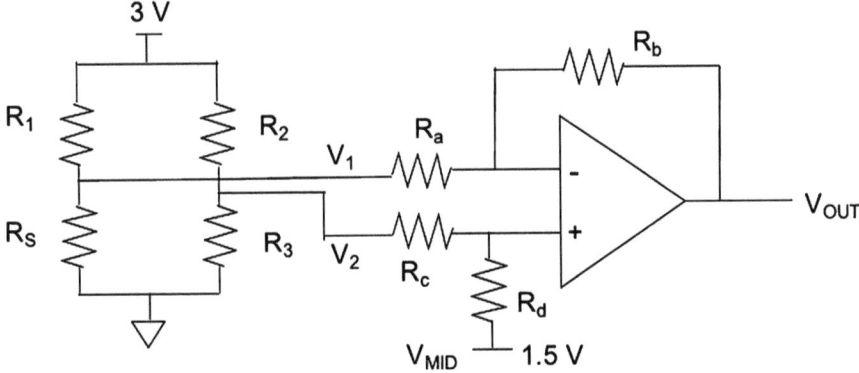

Fig. 3.10 Wheatstone Bridge connected to a differential amplifier

Example 3.1 For a temperature sensor with simple biasing scheme, the sensor temperature varies from 10 kΩ at 0 °C to 9 kΩ at 40 °C. If the supply voltage is 3 V, determine the sensitivity (with the approximate formula) for R_B of 10 and 100 kΩ.

Solution The approximate formula for sensitivity is:

$$\left|\frac{\Delta V}{\Delta R_S}\right| = \frac{R_B V_{DD}}{(R_B + R_{S1})^2}$$

For $R_B = 10$ kΩ, sensitivity is $(10\,k*3)/(20\,k)^2 = 75$ μV/Ω. On the other hand, for $R_B = 100$ kΩ, sensitivity is $(100\,k*3)/(110\,k)^2 = 25$ μV/Ω. Thus, the circuit is more sensitive for $R_B = 10$ kΩ. We can find the corresponding ΔV are 75 and 25 mV.

As mentioned before, the Wheatstone Bridge configuration will require a differential amplifier before the output signal can be fed to the ADC. An example scheme is given in Fig. 3.10.

The output voltage of this circuit can be written as:

$$V_{OUT} = \left(1 + \frac{R_b}{R_a}\right)\left(\frac{R_d}{R_c + R_d}\right)V_2 - \left(\frac{R_b}{R_a}\right)V_1$$

For R_S of the same values as before, and using $R_1 = R_2 = 100$ kΩ, while $R_3 = 9.5$ kΩ, we can calculate $V_1 = 0.26$ V (or $-$ 1.24 V wrt V_{MID}), and $V_{21} = 0.27$ V (or $-$ 1.23 V wrt V_{MID}) and $V_{22} = 0.25$ V (or $-$ 1.25 V wrt V_{MID}), for R_{S1} and R_{S2} respectively. We can now design the differential amplifier for maximum voltage swing of 1.5 V. For $R_a = R_c$ and $R_b = R_d$, the equation simplifies to $R_b = 125\ R_a$. Thus, a possible design is $R_a = R_c = 1$ kΩ, and $R_b = R_d \approx 120$ kΩ (E-12 series resistances).

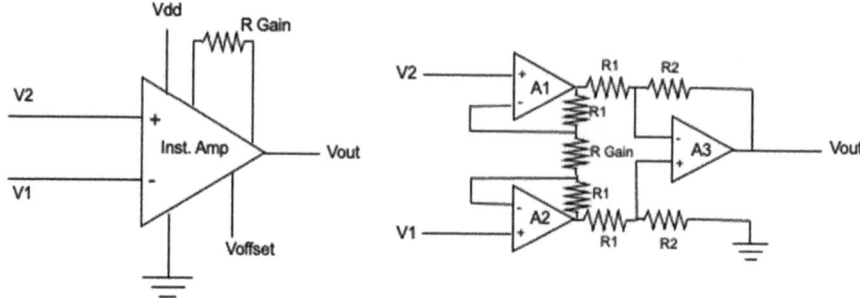

Fig. 3.11 (Left) Instrumentation amplifier circuit example. (Right) Internal op-amp circuit configuration

There is a special type of low-power, low-noise amplifier is common in instrumentation circuits, known as Instrumentation Amplifier (IA). These amplifiers are only used when the inputs are differential and at least one of these conditions are needed to be fulfilled: large gain, large CMRR, low noise, low signal strength, or large input impedance. Also, this offers very small package size. Schematic of IA looks very similar to op-amp, but internally, it is built with 3 op-amps (Fig. 3.11).

The two inputs of the IA are V_1 (negative input) and V_2 (positive input). The differential mode component of the output is: $V_{dm} = G_{dm}(V_1 - V_2)$, where G_{dm} is the differential mode gain. The common mode component of the output is: $V_{cm} = G_{cm}(V_1 + V_2)/2$, where G_{cm} is the common mode gain. The overall output is the sum of the independent common and differential components: $V_{out} = V_{cm} + V_{dm}$. The Common Mode Rejection Ratio (CMRR) is defined (in dB) as:

$$CMRR = 20 \log \frac{G_{dm}}{G_{cm}}$$

A large CMRR indicates that signals present on both V_1 and V_2 (such as coupled noise of the electrode wires) will have little influence on the value of V_{out}. IA is required in many signal acquisition AFE such as electrocardiography (ECG) or electroencephalography (EEG) signal collection.

Some rule-of-thumb op-amp based AFE design rules:

- Match input impedance to improve CMRR.
- Use IA for signals where noise present on both differential inputs needed to be rejected and only differential voltage is needed to be amplified.
- Choose quality components in sensitive sections of AFE. (Note that typical passive component tolerances can be as high as 10%).
- For linear response, a negative feedback amplifier is needed.
- Assume no current is flowing into the inputs of the op-amps (and IA).
- Assume both input terminals of the op-amp (and IA) are at the exact same voltage level (i.e., $V+ = V-$).
- For feedback resistances, keep it within the range of 1kΩ to 1 MΩ.

- Circuit bandwidth (BW) is dependent on the op-amp frequency response and the circuit gain.

While in this section, we have discussed the most common resistive type of analog sensors, there are other types of analog sensors too, such as capacitive sensors, touch sensors, diode-type sensors, or transistor type sensors. For example, most acceleration sensors (e.g., IMU) are capacitive type sensors and fabricated with MEMS technology. All sensors are basically transducers that convert a physical data to electronic signals. With modern technology, sensors can be designed for virtually every physical and chemical quantities, such as weight, velocity, acceleration, electrical current, voltage, temperatures, and chemical compound detection. Physical constructs used for sensor constructions include law of induction (i.e., generation of voltages in an electric field), bio-impedance, bio-electricity, piezo-effect, or light-electric effect. In recent years, huge amounts of sensors are designed and commercially available in the market that you should be able to find off-the-shelf sensors for almost all embedded system projects. Note that although almost all sensors are analog, there are lot of digital sensors available in the market where the housing already includes the required AFE to convert the analog signal to digital data, thus easy to interface with your MCU. Examples of digital sensors (which uses integrated analog transducers) include digital temperature and humidity sensor, proximity sensor, charge coupled device (CCD) camera, and touch sensors. There are some inherently digital sensors (or interface) as well, such as reed switch, CMOS image sensor, and keypad. To interface a digital sensor, you need to connect the interfacing wires with correct digital pins of your MCU and load the corresponding communication library in the firmware. Note the supply voltage ranges of the digital sensors as compared the your MCU, and be mindful that digital logic levels can differ significantly. For instance, a 5 V CMOS logic considers any voltage between 1.5 and 3.5 V as indeterminant, while 3.3 V LVTTL logic considers 3.3 V as Logic 1. Thus, direct interfacing will not work. For these types of cases, you can simply use a digital logic level converter for proper interfacing.

A common bug in interfacing switches with MCU can be caused by mechanical switch bounce. This bouncing is the result of a mechanical switch to close and open several times before finally settling due to its spring action. The result is that the software might record multiple ON/OFF actions for a single switch operation. Bouncing can have duration of 1–10 ms. Expensive high-end switches will have less of the issue. If this issue arises, this can be solved with in hardware with adding a series resistor (e.g., 1 kΩ) followed by a capacitor (e.g., 0.1 uF) before the digital pin of the MCU. It can also be solved in software by adding a 10 ms delay or disarming the interrupt for 10 ms followed by clearing the flag and rearming it, although this reduces foreground efficiency. Other approaches are using Schmitt trigger circuit or blind cycling.

3.4.3 Analog Filters

A common issue with analog signals is inherent noise. Unlike digital signal, analog signal is stochastic in nature with various types of noises. Some of these noises are internal, such as thermal, shot, flicker, burst, and transit-time noise. Some other noises are externally coupled, such as inductive and capacitive.

Thermal noise, also called Johnson-Nyquist noise, is unavailable and is generated due to ambient temperature. Higher temperature leads to higher thermal noise. Shot noise, governed by Schottky formula, is also unavoidable. This noise is generated due to traversing a junction, for example in diode or transistor. Flicker noise, also known as pink noise, is related to DC current and inversely related to frequency. Burst noise is non-Gaussian sudden step-like random noise in the order of 100 μV. Transit mode noise is a high frequency noise at Very High Frequency (VHF) or higher frequencies.

The magnetic noise is coupled from external alternating magnetic fields. As coupled alternating magnetic fields induce alternating currents, especially in long straight traces, this can cause interference in analog signals, especially if the signal is small in magnitude. This magnetic noise can be modeled with mutual inductance (S). The most common cases are magnetically induced noise from utility lines (50 or 60 Hz depending on the country), which can be reduced or eliminated with an appropriately designed notch filter. The electric-field noise, coupled with alternating electric fields inducing alternating voltages due to capacitive coupling, can also be an issue in some systems, especially in mixed-signal circuit designs. This coupling can be modeled with stray capacitance (C). To reduce externally coupled noises, one approach can be to reduce noise source (where applicable) by enclosing it in a grounded metal box, increase the transition time (rise time and fall time) of noisy signals, and limiting dI/dt in the coil. If noise source is not controllable, then the coupling can be reduced in your instrument with a combination of approaches like maximizing distance of the instrument from the noise source, using twisted pair or co-axial cables for sensitive signal carrying wires, using shielding with a shield plane or shielding traces, reduce length of straight wires or traces, place delicate electronics in a grounded case, optical or transformer isolation circuits, and/or using appropriate noise filters.

Analog filters can be active or passive types based on the components. In passive filter, all components are passive (i.e., resistor, capacitor, and inductor), whereas in active filter, at least one of the components is active (such as transistor or op-amp). Passive filters are simple, cheap, and smaller in footprint, but do not perform well. Active filters are superior in performance as well as can provide some gain, although typical designs of these filters keep the gain to 1. Note that in most ES design, inductors are rarely used due it's larger size, cost, and complexity. The common analog filter types based on filter response are Butterworth, Chebyshev, and Bessel. In this section, general descriptions of some common noise filters are provided. Note that as the ES designer, you should be aware that you can design filters in digital domain (software based) using Finite Impulse Response (FIR) or Infinite Impulse

Response (IIR) topologies as well. With regards to ES, key advantages of analog filters are high performance and low-latency, while disadvantage are cost and requirement of certain footprint. On the other hand, the digital filters have the counter advantages and disadvantages.

3.4.3.1 High-Pass Filter (HPF)

A high-pass filter (HPF) will allow higher frequencies than the cutoff (f_{HPF}) to pass, and suppress frequencies lower than f_{HPF}. A passive type of HPF can be made by using a capacitor in series, followed by a resistor to ground. The cutoff frequency of a passive RC filter is given by,

$$f_{HPF} = \frac{1}{2\pi RC}$$

For improved performance, op-amp based active HPF can be utilized. Commonly used active filter design examples in ES include Sallen-Key filters and two-pole active high pass filter with Butterworth response.

3.4.3.2 Low-Pass Filter (LPF)

A low-pass filter (LPF) will allow lower frequencies that the cutoff (f_{LPF}) to pass, while suppressing frequencies higher than f_{LPF}. A passive type LPF can be made by using a resistor in series, followed by a capacitor to ground. The cutoff frequency of this passive RC filter is also given by the same equation as in HPF. Active filter designs can have improved performance, such as Sallen-Key filters, and two-pole active low-pass filter.

3.4.3.3 Band-Pass Filter (BPF)

A band-pass filter (BPF) passes signals of a continuous range of frequencies (f_{BW}) defined by a low cut off (f_L) and a high cut off (f_H). A passive version of this filter can be accomplished by simply concatenating a passive LPF followed by a passive HPF, or vice versa. For active BPF, there are also some standard BPF filter designs, such as Inverting band-pass filter, and Infinite gain multiple feedback active filter. It is common to utilize one of these standard designs in ES.

3.4.3.4   Band-Stop Filter

A band-stop filter is just the inverse of BPF. It is also known as band-reject filter. A BPF can be constructed by parallel combination of HPF and LPF with appropriately designed cutoff frequencies. The use of BPF is rare in ES designs.

3.4.3.5   Notch Filter

Notch filters have very narrow pass-band (or stop band). A very common notch filter with band stop at utility line frequency is found in ES designs. An example of passive notch filter is Narrowband Twin-T Notch Filter, whose cutoff frequency can be calculated as,

$$f_{\text{notch}} = \frac{1}{4\pi RC}$$

This filter design has two parallel branches, one of which is compose of two capacitors (value of C) in series with a parallel resistor (value of R) tapped from the common terminal of the capacitors, while the other is composed of two resistors (value of $2R$) in series with a parallel capacitor (value of $2C$) tapped from the common terminal of the resistors. Another example of passive notch filter is Wideband Notch Filter. An example of commonly used active notch filter is Twin-T Notch Filter with variable Q.

3.4.3.6   Artifact Removal

In addition to noise, artifact might also need to be removed from signals in some cases. The difference between noise and artifact is that artifact are signals from unwanted signal sources. For instance, when recording EEG signal from brain, it might inadvertently record muscle signals originating from eye movements or heart beats. As artifacts are typically in the same spectral range of the signal, it is not possible to use the above filtering approaches (which are spectral filters). Thus, complex statistical approach, wavelet decomposition, or source component analysis might be required [1, 2]. These approaches are beyond the scope of this book. If the noise and artifacts that are common in differential inputs of signals, the use of instrumentation amplifier with large CMRR is an effective approach.

3.4.3.7   Analog Filter Design Examples

Op-amp based analog filter design is the most common type of active filters utilized in ES AFE. There are many standard active analog filter designs available, as mentioned before, which can be utilized for almost all applications. It is typically

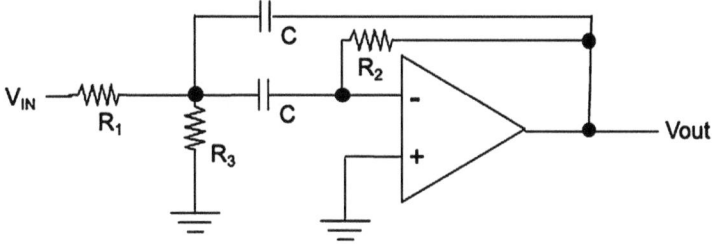

Fig. 3.12 Infinite gain multiple feedback second order active filter design

unnecessary to develop own design. In this section, we discuss design of a band-pass filter as an example.

For this band-pass filter design, let's pick Infinite Gain Multiple Feedback second order active filter. This filter design is given in Fig. 3.12.

The governing equations for this filter are:

$$f_c = \frac{1}{2\pi} \sqrt{\frac{1}{R_1 R_2 C_1 C_2}}$$

$$Q = \frac{f_c}{BW_{3dB}} = \frac{1}{2} \sqrt{\frac{R_1}{R_2}}$$

$$A_v = -\frac{R_2}{2R_1} - 2Q^2$$

Example 3.2 Let's say the target center frequency is $f_c = 1$ kHz. Typical value of A_v is -1 (unity gain, inverting configuration). Design an Infinite Gain Multiple Feedback second order active filter.

Solution Using the above relations, we find, $R_2 = 2R_1$ and $Q = 0.707$. For low-noise amplifier, resistor values should be in the kΩ range. If we take the value of $R_1 = 10$ kΩ, then $R_2 = 20$ kΩ. Thus, using the above equation, we can find the value for $C_1 C_2 = 1.27 \times 10^{-16}$. If we consider, $C_1 = C_2$, then $C_1 = C_2 = 11.27$ nF $\approx$ 10 nF. For R_3, we can use 1 kΩ (does not affect filter performance).

One of the very helpful solution in active filter designs with op-amp is software and online tools, like Analog Filter Wizard from Analog Devices (https://tools. analog.com/en/filterwizard/).

Example 3.3 Using Analog Filter Wizard, design 1 stage Butterworth band-pass filter for (a) low-power and (b) low-noise with the following specifications.

Pass band (-3 dB) $= 100$ Hz. Stop band (-40 dB) $= 20$ kHz. Center frequency $= 1$ kHz. Gain $= 0$ dB, Unipolar 3.7 V supply.

Solution Using the online tool, we will get the following values:

(a) For low-power design: $R_1 = 998 \,\Omega$, $R_2 = 2$ MΩ, $C_1 = C_2 = 1.59$ nF, $R_3 = 5$ kΩ.

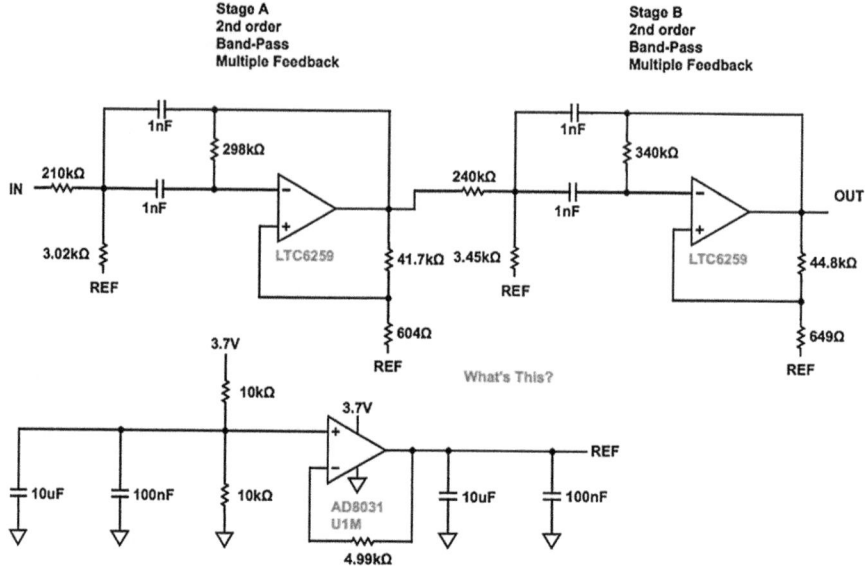

Fig. 3.13 Low-power version of the 2-stage Band-pass filter design (using Analog Filter Wizard)

(b) For low-noise design: $R_1 = 15.9$ kΩ, $R_2 = 31.8$ kΩ, $C_1 = C_2 = 100$ nF, $R_3 = 1$ kΩ.

Note that the online tool will show an op-amp based mid-voltage generator circuit like Fig. 3.8. This is needed for unipolar supply to create the equivalent bipolar supply with respect to mid-voltage (V_{mid}).

Example 3.4 Using Analog Filter Wizard, design a 2-stage band-pass filter with the following specifications:

Center frequency = 5 kHz. Pass band (−3 dB) = 1 kHz. Stop band (−40 dB) = 10 kHz. Gain = 0 dB. +Vs = 3.7 V. −Vs = 0 V. Optimization: (a) Low-power, and (b) Low-noise.

Solution The designed filter for low-power is shown in Fig. 3.13 and low-noise is shown in Fig. 3.14.

Note that the low-power version has higher values resistors (to reduce currents) while lower values of capacitors (to satisfy the filter response). The low-noise version, however, uses lower values of resistors (lesser relative noise). Both circuits include a mid-voltage generator circuit indicated as REF (reference voltage) output.

The wizard also allows generating a list of components, commonly referred to as Bill of Materials (BOM). As most electronic components are fabricated within a certain tolerance, the exact response might shift from ideal response based on the exact values of components used. This wizard also provides an opportunity to find out deviation of response due to common component tolerances. If the deviation is

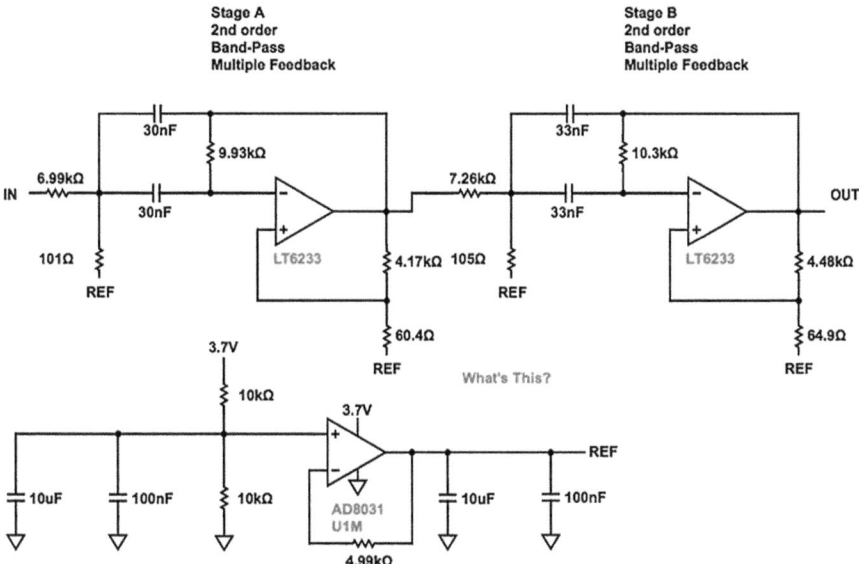

Fig. 3.14 Low-noise version of the 2-stage Band-pass filter design (using Analog Filter Wizard)

not acceptable, the designer can choose components with lesser tolerances (typically leads to higher component cost).

Later in this chapter, we will learn how to take a schematic design to develop your custom PCB layout.

3.4.4 AMux, ADC, and DAC

As microcontroller units (MCU) are digital devices, all analog signal needs to be digitized before MCU can analyze and operate on these data. For this digitization, an Analog to Digital Converter (ADC or A/D) circuit is used. ADC converts analog signal to digital data that internally uses a Digital to Analog Converter (DAC or D/A). DAC is also useful to convert an MCU digital output to analog values. Almost all MCU contain ADC; however, DAC for output is not that common. If you need to convert multiple analog inputs to digital values, instead of using multiple ADCs, you can use a circuit called Analog Multiplexor (AMux) that can sample different analog signals and feed one signal at a time to ADC. This approach reduces hardware expense significantly, and commonly used in most low-cost MCUs. A system composed of multiple AFE with data acquisition from them (via AMux-ADC or multiple ADCs) is called Data Acquisition System (DAS).

To understand how ADC works, it is important to first understand how DAC works. The simplest DAC system is R-2R ladder. Here, a brief discussion is presented from fundamental concept formation. Figure 3.15 shows a 4-bit R-2R

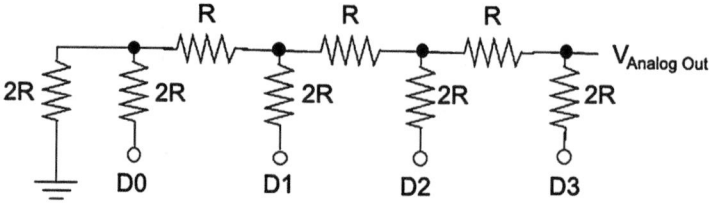

Fig. 3.15 A 4-bit R-2R ladder DAC

ladder DAC. The inputs are represented as "1," "2," "4," and "8," where 1 is the least significant bit (LSB) and 8 is the most significant bit (MSB) of the 4-bit input digital data.

Let us consider an input of 1000_b, which represents 8. This is the mid-level of values that this 4-bit can hold, thus we expect a mid-level voltage output. For ease of calculation, let us assume a 5 V system (i.e., 5 V represents Logic 1, as well as 5 V is the highest output level). For the input 1000_b, the input "D3" should be at 5 V, whereas all other inputs are at 0 V. To analyze the circuit, let's consider the two leftmost resistors (connected to Gnd and D0), which are 2R each. Thus, their parallel equivalent is R. This equivalent R is in series with the left horizontal resistor R, making a 2R resistor. This 2R combines in parallel to the 2R resistor of input "D1." We can continue this reasoning until the last series R connected to the output. Now, for the input "D3" which is connected to 5 V, the circuit becomes a simple voltage divider with 2R on each branch and output connected to the middle. Thus, the output voltage will be 2.5 V, which is the needed analog voltage. This analysis technique is based on Thevenin Theorem.

Example 3.5 For the 4-bit R-2R ladder (5 V system), find the analog output value when the digital input is 0001_b.

Solution We will utilize Thevenin Theorem to analyze this circuit. Here, the inputs $D0 = 5$ V, and $D1 = D2 = D3 = 0$ V. The leftmost 2R-2R acts as a voltage divider with output of 2.5 V. For the next stage, the equivalent resistor with the 2.5 V supply is $2R \| 2R + R = 2R$, and another 2R is connect to 0 V (input D1). Thus, the output voltage is 1.25 V. Using similar argument, the output voltage at the input stage of D2 is 0.625 V. Finally, the output voltage of last stage is 0.3125 V. For cross-checking, as the input represents 1/16th of the available output, thus 5 V/16 = 0.3125 V.

This is also the smallest value that the DAC can generate, as well as the smallest increment between any subsequent values. Thus, this is called quantization step. In other words, this 4-bit DAC with a 5 V system cannot produce voltage increments in any finer values than 0.3125 V (quantization step).

Example 3.6 For the 4-bit R-2R ladder (5 V system), find the analog output value when the digital input is 1010_b.

Solution Again, utilizing Thevenin Theorem along with Superposition Theorem, the inputs of Fig. 3.15 are: $D0 = 0$ V, $D1 = 5$ V, $D2 = 0$ V, and $D3 = 5$ V.

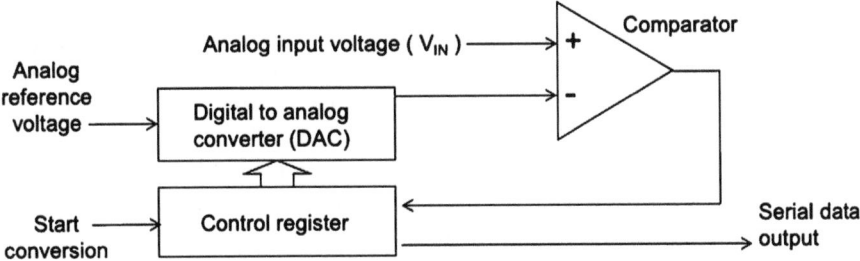

Fig. 3.16 A successive approximation ADC configuration

Considering only input D1 and replacing input D3 with 0 V, we use the similar argument as in the last example to find out that the output will be 0.625 V. Now, considering only input D3 and replacing input D1 with 0 V, we will find the output to be 2.5 V. Now, we can apply Superposition Theorem to consider both inputs to be present, which will result in both outputs to be added, producing 3.125 V at the output for the given input digital data.

In general, the output of a DAC can be computed using the following expression:

$$V_{OUT} = \frac{D}{2^N} V_{DD}$$

where N is the bit-width of the DAC. For this discussed DAC above, $V_{DD} = 5$ V, $N = 4$, and D represents the input digital data. For input of $D = 0001_b$, the output voltage $V_{OUT} = 0.3125$. For input of $D = 1010_b$, the output voltage $V_{OUT} = 3.125$ V. The maximum output voltage of this DAC is $V_{OUT} = 4.6875$ V for $D = 1111_b$. This general expression not only is applicable to R-2R ladder DAC, but to any DAC although other DACs can be more complicated.

As evident from this discussion that the DAC output voltage changes only in a fixed increment. In this example DAC of 4-bits with 5 V supply, this fixed increment was 0.3155 V. This is called resolution (or quantization value). The number of quantization levels in this DAC of 4-bits is $2^N = 2^4 = 16$ levels. We can easily find the resolution for this DAC by $V_{DD}/2^N = 5$ V/$2^4 = 0.3125$ V. The higher number of bits of DAC, the lower resolution value will be, i.e., the DAC will represent the intended analog output in more precise manner.

Some other examples of DAC implementations are Binary weighted DAC (where one resistor is used for each bit of output), Thermometer coded DAC (where resistor for each possible output values are used), Segmented DAC (where thermometer is used for most significant bits, and binary weighted are used for least significant bits), and Hybrid DAC (that uses a combination of other techniques). In addition, some MCUs provide pulse width modulator (PWM) which can be utilized to convert a digital value to analog output, by utilizing a proper low-pass output filter.

ADCs utilize DAC internally for operation. Here we will discuss a simple but commonly used ADC configuration, known as "Successive Approximation." The general block diagram of this type of ADC is shown in Fig. 3.16.

This ADC takes in an analog input signal (to be digitized) as one input (positive terminal) of a comparator. The other input (negative terminal) of the comparator comes from the output of a DAC. The output of the comparator is fed into a digital controller which contains successive approximation register (SAR). The digital output of this SAR is fed to the DAC as digital inputs. The DAC uses an analog reference signal, which is typically V_{DD} of the system. Initially the SAR output is set to 1 for MSB, and 0 for rest of the bits. If the comparator output is positive (i.e., comparator positive input is higher than the negative input), this bit value of 1 in SAR is kept, otherwise it is set to 0. Then the next highest bit is set to 1 and the process continues until all bits are determined. Because this process continues in successive order for all the bits, this ADC is called successive approximation.

Example 3.7 For a 4-bit successive approximation ADC (5 V system), find the digital output when the analog input is 3.3 V.

Solution In this example, this ADC is 4-bit (i.e., the DAC inside the ADC and SAR are 4-bits). Also given that the input voltage is 3.3 V, while the supply voltage (V_{DD}) is 5 V. When the start conversion signal is sent to the controller, it will set the SAR value of 1000_b. The DAC will convert this digital value to 2.5 V. When this 2.5 V is compared with analog input of 3.3 V, the comparator output will be positive (as the input 3.3 V connected to positive input terminal is higher than the 2.5 V connected to negative input terminal). This positive output of the comparator will signal to controller to keep the SAR bit of 1 at MSB. The controller will set the SAR value for next round of approximation to 1100_b. The DAC will produce 3.75 V. As input voltage (3.3 V) is lower than this DAC output, the comparator output will be negative. Thus, the controller will correct the new bit to 0, and will set the next bit to 1 for the next round. Thus, new SAR value will be 1010_b. The DAC output for this value will be 3.125 V. As this is lower than the analog input value of 3.3 V, the comparator output will be positive and the new bit of 1 will be kept. Lastly, the controller will set the next bit (i.e., LSB) to 1. Thus, the SAR value will be 1011_b. The corresponding output voltage of DAC is 3.4375 V. As this is higher than the input voltage of 3.3 V, the comparator will produce negative output, and the controller will correct the SAR value to 1010_b. After 4 approximation steps, this 4-bit ADC completes computation of the digital value, and it will send this computed digital value of 1010_b to the output port.

Note that this digital value corresponds to 3.125 V, which is not exactly 3.3 V, but this is the best approximation this 4-bit ADC can do for this input. Thus, there remains an error, which is called "Quantization Error." If the ADC had higher number of bits, it could have reduced this error. In other words, higher number of bits for ADC will have lower quantization error. The average value of this quantization error is given by $V_{DD}/2^N$ where N is the number of bits of ADC.

Another popular ADC configuration is Sigma–Delta ADC. It contains an integrator, a DAC, a comparator, and a summing junction. It is often used in digital multimeters, panel meters, and data acquisition boards. This ADC is relatively inexpensive and requires only a few external components. But they can obtain high-resolution measurements using oversampling techniques. However, it works best with low-bandwidth signals (a few kHz), it typically has better noise rejection than many others. The advantages of this type of ADC are high resolution, high integration, low-power consumption, and low cost. The disadvantages include slow operation, requirement of a high-speed clock, and not suitable for multiplexing. Another ADC configuration that is found commonly is dual-slope ADC.

ADC requires certain amount of time to compute the digital value, thus the input analog value must be kept constant during this time. This is done with a simple hardware known as "Sample and Hold" (S/H) circuit. This circuit can be simply considered as a capacitor with a series switch. When the sampling occurs, the switch closes momentarily, then opens. This charges the capacitor to the input analog value of that moment and keeps this analog voltage until the ADC computation is complete. When a new data needs to be computed, the switch closes momentarily again to sample the analog voltage, and so on. The output of the S/H circuit is connected to the analog input voltage shown in Fig. 3.16. In most practical implementations, the switch is realized with a transistor (e.g., MOSFET) based switch and typically switched with a periodic clock circuit which determines the ADC sampling rate. The S/H circuit transforms the analog input signal, $e(t)$, to a discrete sequence, $h(t)$, of impulses. The sampling of a signal must satisfy two conditions: (1) the signal is band-limited, and (2) sampling intervals are small such that there is no overlap. This transformation changes the signal from time domain to value domain. In this transformation, the sampling rate must be twice or more of the maximum frequency in the analog signal. This requirement of sampling rate is called "Nyquist Criterion." Frequency satisfying Nyquist criterion is known as Nyquist sampling frequency ($f_{Nyquist}$). Mathematically, $f_{Nyquist} \geq 2 f_{signal}$, which is known as Nyquist limit.

However, in most practical cases, the signals are sampled at a higher rate than this Nyquist limit. The technique to sample at a higher rate is known as oversampling. Mathematically, $f_{oversampling} = \beta f_{Nyquist}$, where β is the oversampling factor. Commonly used value for β is 10. Using oversampling technique reduces the need for anti-aliasing filter. Anti-aliasing filter is required if the sampling rate does not meet the Nyquist criterion. However, even if Nyquist criterion is met but oversampling rate is small, it is recommended to use anti-aliasing filter.

Example 3.8 If a signal has the highest frequency content of 1 kHz, what is the Nyquist rate and what is oversampling rate for $\beta = 10$.

Solution The highest frequency of signal, $f_{signal} = 1$ kHz. Thus, Nyquist rate is $f_{Nyquist} = 2 f_{signal} = 2$ kHz. For $\beta = 10$, the oversampling frequency, $f_{oversampling} = \beta f_{Nyquist} = 20$ kHz.

As ADC is a relatively expensive hardware in MCU, instead of including multiple ADC inside the MCU, some designs will include an Analog Multiplexor (AMux). This M-to-1 AMux will sample each of M inputs one-by-one and feed to the ADC input. Thus, the programmer can convert M analog inputs with only one ADC. Typical value of M is 8. Internally a controller hardware will determine which inputs of AMux to be enabled and fed to the ADC in sequence, and store the output of the

ADC in the corresponding output register of the ADC, so that AMux operation is transparent to the programmer.

Multichannel Data Acquisition System (DAS) also contains separate analog front end (AFE) for each analog inputs. The AFE typically consists of amplifier and filter. These analog signals go through an AMux followed by S/H circuit and ADC. A controller hardware selects AMux channel, and puts the ADC output in the correct output register.

Some commonly used parameters and metrics of ADC and DAC are listed below:

Precision: The number of bits of ADC or DAC. Precision had direct implication on quantization levels and errors.

Range: The difference of maximum (V_{RH}) and minimum (V_{RL}) analog values allowed as input or output voltages for ADC or DAC, respectively. Sometimes it is referred as Dynamic Range or Full-scale (FS).

Quantization: It represents the number of discrete levels the analog voltage is divided between V_{RH} and V_{RL}. For an ADC or DAC of N-bits, the number of levels is 2^N.

Resolution: It is the smallest increment value of analog domain. The ADC or DAC cannot resolve values smaller than 1 resolution. The resolution can be expressed as $(V_{RH} - V_{RL})/2^N$.

Sampling rate: The sampling rate for ADC is the rate at which S/H circuit samples the input data. This is the same rate at which ADC will produce digital data.

Data rate: This is the output rate required for an ADC to operate properly. This can be computed as: Data rate, $d = f_{sampling} N$, where $f_{sampling}$ is the signal sampling rate.

Acquisition time: It is the interval between the release of the hold state and the instant when the voltage on the sampling capacitor settles to within 1 LSB.

Maximum Sampling rate: This is maximum rate the ADC can process samples.

Aliasing: This occurs when the input signal frequencies exceed the Nyquist frequency, which produces "aliased" waveforms.

Accuracy: This measures the difference between the actual output and ideal output. Mathematically, accuracy = (actual output – ideal output)/ideal output.

Monotonicity: This property indicates that the output moves in one direction.

Total harmonic distortion (THD): This is the total distortion produced by an ADC or DAC in terms of harmonic distortion, which adds to the noise (i.e., total noise = THD + N, where N is the noise).

Aperture Delay: It is the interval between the sampling edge of the clock signal and the instant when the sampling is taken.

Aperture Jitter: This is the sample-to-sample variation in the aperture delay.

Integral non-linearity (INL): This property of DAC indicates the deviation between the ideal output and the actual output value.

Differential non-linearity (DNL): This property of DAC is the deviation between two analog values corresponding to adjacent input digital values.

Quantization noise: This is the average noise produced by ADC or DAC, which is the same value as the resolution of the ADC or DAC. This is also known as Absolute noise.

Relative noise: This noise is based on signal-to-noise ratio (SNR), which is the ratio of the dynamic range divided by the amount of absolute noise in the system. It is usually expressed in dB unit.

Example 3.9   For an ADC of 10-bits with maximum and minimum voltage range of +5 and −5 V, respectively, find the quantization noise and relative noise.

Solution   The absolute quantization noise $= (V_{RH} - V_{RL})/2^N = (5 + 5)/2^{10} = 9.766$ mV.

The relative noise, SNR $= (5 + 5)$ V/9.766 mV $= 1024$. In dB unit, the relative noise $= 20 \log_{10}(\text{SNR}) = 20 \log_{10}(1024) = 60.2$ dB.

When considering sinusoidal analog voltages, the relative noise can be best expressed with the following equation:

$$Relative\ quantization\ noise = N * 6.02 + 1.76\ dB$$

Here, N is the number of bits of ADC. Using this equation, we can calculate the relative noise of 10-bit, 16-bit, and 24-bit ADCs are 61.96, 98.08, and 146.24 dB, respectively. In reality, noise is higher due to non-ideal behavior of the converters such as non-linearity.

In some special cases, the analog input can be connected without an ADC when only binary decision is required at a certain value (that can be determined by the hardware). In this case, a simple comparator circuit can be used where the analog input can be connected to the positive input terminal, while the negative input terminal can be connected to a reference voltage that acts as the threshold level. If the input analog value is higher than the threshold, the output of the comparator is positive (logic 1), otherwise negative or zero (logic 0). This logic values can be captured using a digital input pin of the MCU.

3.5   Output and Actuators

In ES systems, there are number of peripherals are commonly found. These can be data storage, output devices, or actuators. Some common examples of these devices are briefly mentioned here.

3.5.1   Data Storage

Almost all new generation MCU contain onboard memories in the form of Random Access Memory (RAM) and Read Only Memory (ROM). They are volatile and non-volatile, respectively. External memory like Flash chip, MicroSD card, even Hard Disk Drive (HDD) can be incorporated, if needed. However, it would be best to select an MCU with sufficient onboard memory, if possible. Incorporation of MicroSD and HDD will require a file system (such as File Allocation Table, FAT) that makes the overall data storage process slow, complex, and requires overhead. Also, this type of ES with file system will also require operating system (OS). On the other hand, advantages of MicroSD and HDD are the possibility of removing from the system for external analysis, very large capacity, and ease of upgrade. The flash chips are non-removable, but fast in operation and lower in power consumption.

3.5.2   Various Analog and Digital Actuators

Actuators of ES can be either digital or analog types. Examples of digital actuators include relays, stepper motors, and buzzers. Examples of analog actuators include speakers, and DC motors. Actuators can be a combination as well, such as robotic arm drive mechanism. Furthermore, there are a lot of display systems utilized in ES. Examples of these are LEDs, seven segment display, LCDs, and graphical display monitors.

Some actuators might require special drive circuitry. This is specially true for actuators requiring higher currents than MCU can drive. Typical MCUs (e.g., Arduino) can drive up to 40 mA circuits. If circuits require higher current, driver circuits must be used. Digital drive circuit can be as simple as relay, whereas analog driver circuit might be simple such as power amplifier, or complex such as motor drive circuits. When using actuators that involve magnetic inductor (e.g., relays), ensure to use a back EMF bypassing diode across the actuator.

3.5.3   Output Drive Circuitry and Ports

When connecting drive circuitry, note that some MCU output requires external pull-up resistors. Carefully study the datasheet of MCU to find out this requirement as well as maximum current ratings. Also pay special attention to available I/O (input/output) ports. These ports can be analog or digital. Analog ports are typically connected to ADC or DAC. Digital ports can be General Purpose I/O (GPIO), PWM, serial port, parallel port, USB, Ethernet, I2C (or I^2C), SPI, SCI, VGA, or COM ports.

3.5.4 Serial Communication

When using the serial transmission technique, the bits making up the data characters are transmitted one-at-a-time using a single transmission channel. At the receiving end, the serially received bits are reassembled into their original parallel form. Data is transmitted in a sequence of 1 s (called "Mark") and 0 s (called "Space"). An example of a serial port is RS-232 which is a non-return-to-zero (NRZ) format port. This also requires a ground wire that allows DC current to flow through it, and has a slightly different value at each end of the interface. If nothing is transmitted (idle mode), the channel will be kept in a "Mark Hold" state. The transmission begins with a "Start" bit which switches from the Mark Hold state to the Space state. The start bit is the first bit used in the "Framing" of the data character. This is considered to be an overhead bit since it is not part of the data character that will be transmitted.

The serial port is commonly found in ES hardware in the form of Universal Asynchronous Receiver-Transmitter (UART or USART). The transmitter hardware includes a register to hold the outgoing serial bit stream. Typical data size is 8-bit (i.e., 1 Byte) that is written to the transmission buffer (Tx Buffer). Data is passed through this First-in-First-out (FIFO) buffer and reaches the transmission shift register. This register adds three more bits: Start bit, Stop bit, and Parity bit. The parity bit serves as a quick check if the data becomes corrupted at the receiver end. In addition, the hardware contains some transmitter flags, such as transmission ready (TRDY), Transmission Overrun Error (TOE), and Transmitter Shift Register Empty (TMT).

The receiver hardware similarly consists or a receive shift register, where start and stop bits are discarded. The parity bit is used to ensure data integrity. The 8-bit data is passed through the receive buffer (Rx Buffer) and becomes available to ready by the MCU software code when it passes through this FIFO buffer. A set of receive flags are also set accordingly, such as Receive Ready (RRDY), Receive Overrun Error (ROE), Stop detection bit error (BRK), Frame Error (FE), and Parity Error (PE).

UART is a common communication protocol between a computer (referred to as Data terminal equipment, DTE) and a modem or printer (referred to as Data communication equipment, DCE). UART port is bi-directional (duplex) and needs to be connected in cross manner, i.e., TX pin of transmitter to be connected to RX pin of receiver, and vice versa. In duplex communication, both parties can transmit and receive, whereas in simplex communication, only one party can transmit and the other can only receive. There also handshaking pins such as RTS and CTS, which needs to be cross connected as well. The transmitter RTS becomes 1 when the data transmission is occurring signaling the receiver CTS that data is being sent. However, UART works without these handshaking pins as well, if the number of wires needs to be minimized (although not recommended). Also, it is a good idea to insert a small resistance in series to TX and RX pins to avoid damage from accidental short-circuits.

This type of serial port is also used in Serial Communication Interface (SCI) protocol. DB9 port of computers uses SCI communication. A driver chip (e.g.,

MAX232A) is used as a drive circuit in this type of port. SCI port can be used to connect multiple computers in half-duplex mode with a shared serial bus. In half-duplex system, any party can transmit or receive, but only one is allowed at a certain time. SCI is an asynchronous communication protocol, where clock signal is not shared between the parties.

Serial Peripheral Interfacing (SPI) is a synchronous protocol with a shared clock (from master device to slave device). Synchronous allows a faster communication, but it is also expensive compared to asynchronous. SPI port has four pins: SCLK (serial clock), MOSI (Master Out Slave In), MISO (Master In Slave Out), and SS (Slave Select). The SCLK is produced by the master device and fed to all slave devices (synchronous). The MOSI transmits data from master to slave, whereas MISO transmits data from slave to master. The SS pin indicates if the slave needs to be active. This SS pin typically becomes active when low (active low). The register is typically 8-bit (1 Byte) and denoted as SPDR. When data needs to be transmitted, this register is written with the data then communication is started. When receiving data, the same register will hold the received data. SPI can connect to multiple slaves with one master device. This can be done in two configurations: Independent SPI or Daisy Chain SPI.

In Independent SPI, SCLK, MOSI, and MISO pins are tied together for all master and slaves. The Master has dedicated SS pins for each slave. When the master communicates with a particular slave device, first it turns on the corresponding slave using the proper SS pin. All other SS pin must be in disabled state, otherwise data will be corrupted. Note that typical SS pins in MCUs are active low.

In Daisy Chain SPI, the SCLK pins are shared, but MOSI/MISO pins are connected in sequence like in series. It requires only one SS pin, thus reducing hardware cost. When data is transmitted to a particular slave, the clock is provided such that the transmitted data from master passes through other slave devices in between the master and the target slave device.

Another serial communication port is Inter-Integrated Circuit (I2C or I^2C). This is a multi-master, multi-slave, single-ended, serial computer bus invented by Philips Semiconductors (now NXP Semiconductors). This uses 2 wires only: SDA (Serial Data Line) and SCL (Serial Clock Line). These lines must be pulled up by appropriate pull-up register, whose value depends on the number of devices connected to this bus. The protocol dictates that after the Start bit (indicated by a low transition of SDA line), address needs to be transmitted first. The device with the address will become active for the remaining transmission until Stop bit is sent. Data is transmitted in 8-bit (1 Byte) format, after which acknowledgment (Ack) is received. The data transmission rate is typically 100–400 kHz.

Another very popular serial communication protocol is Universal Serial Bus (USB). USB is invented in 1994 by a group of seven companies: Compaq, DEC, IBM, Intel, Nortel, Microsoft, and NEC. There are three basic formats of USB:

1. *Standard:* This format is intended for desktop or portable equipment.
2. *Mini:* This format is intended for mobile equipment.
3. *Micro:* This is a low-profile format for mobile equipment.

Each of these have two types of ports: Type A and Type B. However, newer generations have been developed that surpassed these original formats such as USB-3, and USB-C (aka Type-C).

Version 1.0 of USB released in 1996 was rated for 1.5 Mbps data rate and 5 V, 1.5 A power ratings. Version 2.0 released in 2000 improved this to 480 Mbps with 20 V, 5 A power option. The version 3.0 released in 2008 further improved this to 6 Gbps data rate and 20 V, 5 A power ratings. This is sufficient for many applications; thus, USB has seen wide adoption in different devices for both data communication and charging. USB uses four wires (& shield): Vdd (+5 V DC power line, Red wire), Gnd (0 V ground, Black wire), Data+ (Direct data, Green wire), and Data- (Inverted data, White wire).

Serial communication encoding can be various types, such as digital (positive logic, unipolar, used in digital logic circuits), NRZ (Negative logic, bipolar, used in RS232 port), NRZI (transition represents logic 0, no transition represents logic 1, used in USB), and Manchester (positive slope transition represents logic 1, negative slope transition represents logic 0, used in Ethernet).

3.5.5 Parallel Communication

Parallel ports can allow multiple bits or symbols to be transmitted as a group at the same time. It increases throughput of the port, but it requires more hardware thus relatively expensive. Examples of parallel port are dedicated printer ports (e.g., LPT or DB25), disk drive, and graphical display port. Most of these parallel ports were designed because serial port was not fast enough for the required throughput. But with newer high-speed USB, some of these now can be handled by USB such as printer port. In ES, some other display components such as LCD or 7-segment display also might need parallel port interfacing. In some parallel port interfacing, driver circuits might be required. For example, when interfacing a stepper motor with a MCU, a motor driver interfacing is required to process the parallel port data from the MCU and provide sufficient current to each coil of the stepper motor.

Inputs can also be parallel port, for example an array of switches. Each switch will connect to a specific input pin so that every combination can be recognized. However, this requires $n + 1$ wires for n-switches. This is the simplest scheme used when multiple simultaneous switch activations are required. However, this might not be feasible if the number of switches is large, for example a 4×4 keypad. Dedicating 16 input pins for this keypad interface might not be possible in many MCU units, and not efficient for others as well. To resolve this problem, a solution is to implement scanned interfacing. In this interfacing, the switches are arranged in a matrix or grid with rows (one pin of the switches) are connected to output ports of the MCU, while columns (other pin of the switches) are connected to input ports of the MCU. If the output ports are open collector, then each input port will require pull-up resistors. The output ports are energized one at a time in a round-robin fashion. In one scan, if a switch is pressed, only one input pin will be different (corresponding to

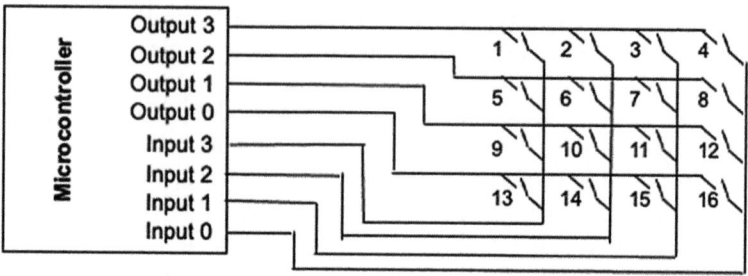

Fig. 3.17 A 4 × 4 scanned interfacing example

the row and column of the switch pressed) than all others. This can be easily decoded using software to determine which switch was pressed. A 4 × 4 scanned interfacing example is shown in Fig. 3.17.

In this example, due to the open collector output, the input ports are pulled to HIGH when no switch is pressed. When a switch is pressed, the corresponding column input becomes LOW only when that corresponding row is enabled through the scanning process. Thus, in this example, only 8 pins (4 outputs and 4 inputs) are sufficient to scan 16 switches. The advantage becomes more prominent when larger switch matrix is considered.

One limitation of this technique is that the input speed must be sufficiently low such that only one switch is ON during one scan. This is typically not an issue when the switches are pressed by human. Another caution must be noted that if the switch is pressed long enough for multiple scan cycles, it should be misinterpreted as multiple keystrokes. This can be resolved by clever programming.

In a more extreme case with a very large number of keys, such as 16 × 16 = 256 keyboard, the number of input output pins might still be too many to accommodate in an MCU. In these cases, multiplexed scanned interfacing can be utilized where a set of multiplexor (mux) and de-multiplexor (demux) can be used. In this example of 16 × 16 matrix, a demux of 4-to-16 can be used to collect to 4 output pins that can produce 16 outputs in sequence (round-robin fashion). These 16 outputs of demux can be fed to rows of the matrix, connecting one sides of the switches in each row. The other sides of the switches from each column can be connected to the 16-to-4 mux. The 4-pin output of this mux is connected to 4 input pins of the MCU. Thus, only 8 pins of the MCU were needed to connect this 256-switch keyboard!

Scanned interfacing technique can also be used for output interfacing such as LED matrix. For example, a 4-pin output ports can enable transistor switches to 4-rows of LEDs. If each row has 6 LEDs, another 6 output ports can be connected to another set of transistor switches for each column. Thus only 10 output pins are sufficient to drive 24 LED matrix.

3.6 Peripheral and Bus System

As MCUs are computers-in-a-chip, they integrate bus and peripherals. The bus system connects the microprocessor with memory and input/output (I/O) ports. It is important for ES designers to understand how these ports work to be able to efficiently utilize them in the software program.

3.6.1 Buffer Queue

Each I/O port, whether input or output, contains some hardware buffer. These are typically First-In-First-Out (FIFO) type and small-in-size. Thus, a software code must serve the port (collet data from output port or feed data to input port) sufficiently quickly. If data is not collected from an input port quickly enough, the FIFO might overflow leading to an overflow error. If data is not provided to an output port, then some synchronous communication channel (e.g., BAUD rate-based communication) might transmit incorrect data. Although this can be avoided with other mechanism (such as hardware or control register settings), so here we focus on input port buffer. There are two primary techniques to serve the input buffer queue: Polling and Interrupt.

3.6.1.1 Polling

Polling mechanism is a periodic checking of input port data and fetch the data if there is any new data available (Fig. 3.18). A simple example of polling is to fetch one new data in every loop of a forever-loop code. A little more advanced mechanism is to check the port data ready flag once in loop to see if there is new data. If there is new data available (based on the flag status), then the software code can read this data from the port. While this technique is simple and widely used by new ES programmers, this is not efficient. There are several major drawbacks of this approach. One drawback is the polling rate might not be fast enough to guarantee of no overflow error. Toward this, an ES designer might be tempted to have fast polling rate. But on the other hand, if there is no data at the port, the clock cycles spent to check the port data are wasted. If there are a lot of missed read polling attempts, that leads to a large wasted clock cycle making the system inefficient, not only in terms of program operation but also in terms of energy wasted. This technique especially unsuitable for inputs with sporadic episodes.

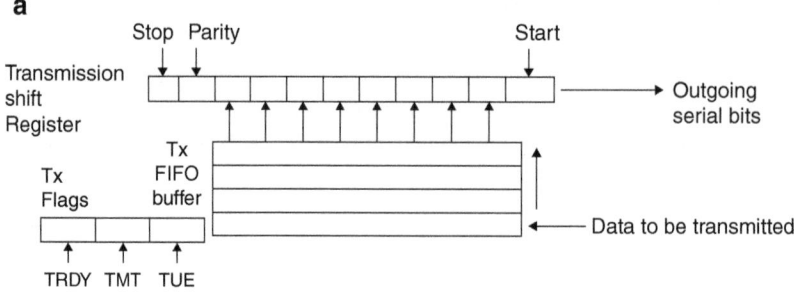

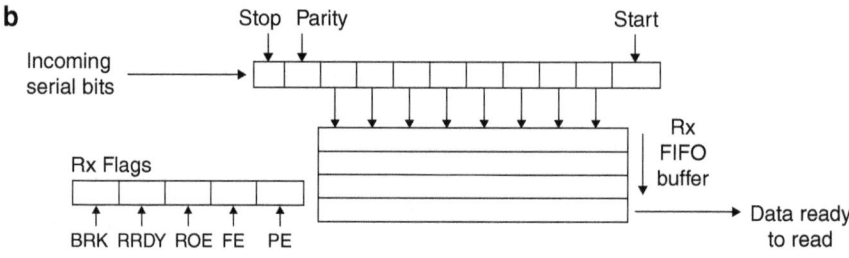

Fig. 3.18 (**a**) Transmitter hardware. (**b**) Receiver hardware

3.6.1.2   Interrupt

This technique is a hardware/software technique and requires deep knowledge and skills of ES designers to properly implement. If implemented properly, this is the most efficient technique for queue mechanism and makes the ES reactive to inputs. To use interrupt in an MCU, the MCU must have the required hardware (an example shown in Fig. 3.19). Study the MCU datasheet to find out the type and number of interrupt available. If MCU is equipped with interrupt, this is the most efficient mechanism to collect received data as it does not waste any clock cycles for checking the port when there is no data. This technique is especially suitable for sporadic inputs. As there are typically only a few interrupt is available in MCU, ES designer must prudently decide which inputs to be checked with interrupt and use polling for the rest.

Interrupt hardware consists of an input driven service request flag. This generates an interrupt request (IRQ) that passes through an AND gate. The other input of the AND gate is connected to register of Interrupt Enable (iEnable). From the MCU firmware, these interrupts can be enabled or disabled by setting or resetting the iEnable bits. The output of the AND gate is connected to Interrupt Pending (iPending) register. This indicates which interrupt has occurred. All iPending bits are ORed together and fed to another AND gate, whose other input is connected to Global Interrupt Enable (PIE) bit. The output of this AND gate serves as the interrupt

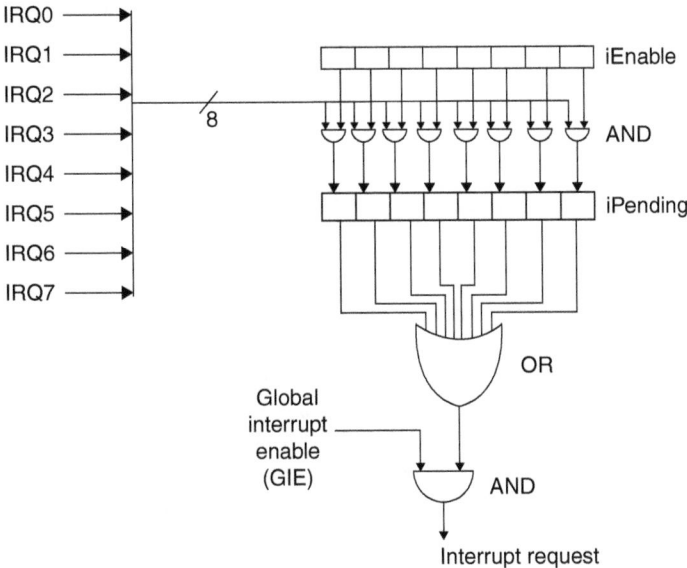

Fig. 3.19 An example of interrupt hardware

request to the MCU. By setting the PIE bit to 0, all interrupts can be disabled. By default the MCU in Arduino PIE is 0.

The peripheral must be instantiated with hardware capability in order to generate an interrupt request. Each peripheral is assigned its own interrupt request number. The interrupt system follows a sequence of procedure:

1. The initialization software (e.g., setup() of Arduino) for that peripheral must set the appropriate bits in the peripheral IRQ enable register (iEnable in the figure).
2. Hardware interrupt request lines (IRQ) bring the different peripheral interrupt requests into the MCU where they are masked (using AND gates) by the iEnable control register. In order for a particular interrupt request to get through, the appropriate iEnable bit must be set by software when the system is initialized.
3. Interrupt requests that get through the AND gates, apply 1's to the corresponding bit in the iPending control register (as shown in the figure).
4. The iPending bits are then OR-ed together to generate a processor interrupt request which may be masked (AND gate) by the PIE (or GIE) bit located in the status control register. The PIE (or GIE) bit must be set by software when the system is initialized.
5. This output of this mask AND gate is the interrupt request that is fed to the MCU control unit.

Interrupt causes exception to uP operation to service any interrupt. Before uP can service the interrupt, if the uP was performing any operation, all register content must be saved in the stack memory. The interrupt service is done through Interrupt

Handler that performs Interrupt Service Routine (ISR). After service is complete, all uP register contents are restored back from the stack memory, and the uP continues its normal operation.

3.6.2   Direct Memory Access (DMA)

In case of high-speed I/O operation, such as reading or writing of disk, flash memory, or microSD card, simple queueing technique (either polling or interrupt) is insufficient. One of the major issues is that if the peripheral I/O port is sending/receiving large amount of data (for example a large buffer content, a data file, or a high-speed ADC data), the queueing technique requires the data to be read from the port to uP register (load) then write from uP register to memory (store), or to be read from the memory to uP register (load) then write from uP register to the port (store). Each of these load-store cycles occupies the bus between the microprocessor (uP), port, and memory. Not only this is a slow process for large repetitive data, but also the uP is performing this load-store commands and is not able to perform other tasks, if needed. In addition, as the uP is actively involved with load-store, the uP cannot be put in a sleep mode to save energy.

To resolve this type of situation, which are commonly in ES systems, a method known as Direct Memory Access (DMA) is available in some MCU. DMA controller can independently read from the input port, or write to output port without the processor getting involved (Fig. 3.20a, b). In the DMA read operation cycles, data from the input port goes to predefined location of the memory (RAM) without going to uP register. Similarly in the DMA write operation cycles, data from predefined location of memory (RAM) goes to the output port. Note that all MCU, especially low-end inexpensive MCUs might not have DMA. For example, Arduino Uno MCU Microchip ATmega328P does not have DMA, but Arduino Due MCU Atmel SAM3X8E ARM Cortex-M has DMA. DMA controllers can be programmed to perform certain amount of data transfer as per program. After the amount of data has been transferred, DMA can restart or give control back to uP.

There are various types of DMA setup (Fig. 3.20c, d). One of the common types is Burst, where the desired I/O bandwidth matches the computer bus bandwidth. That means the data transfer happens at the maximum allowed rate. All other operation of the bus is halted. The block of data is transferred at once without interruption. This should be used for high-priority operations. An example of this type can be when a large file data needs to be transferred. Another common type is Cycle-steal. In this DMA setup, the I/O bandwidth is less than the computer bus bandwidth. DMA hardware steals cycles when bus is idle. This allows normal bus operation, thus software can run, although slowly. This can be used for low-priority data transfer. An example of this can be Bluetooth data transfer which requires a small buffer to be sent to the serial port. We will discuss further details in the software chapter.

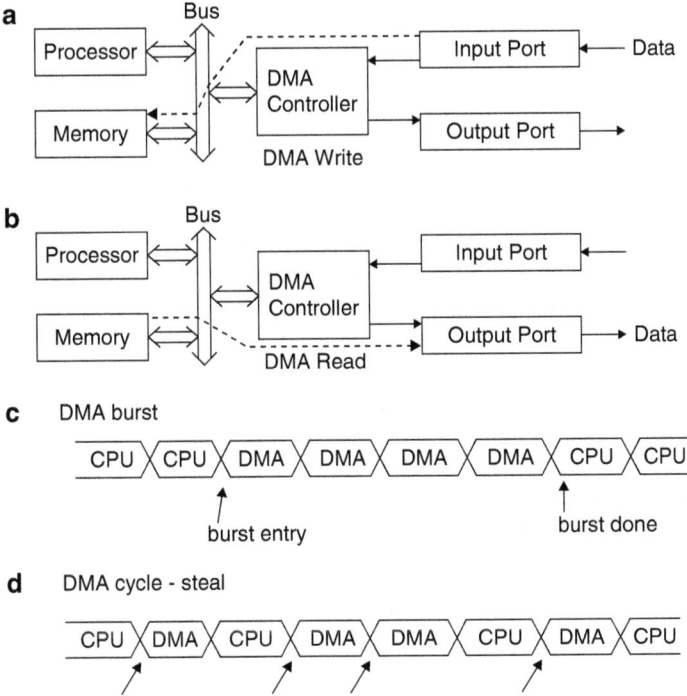

Fig. 3.20 (**a**) DMA read operation. (**b**) DMA write operation. (**c**) DMA burst timing diagram. (**d**) DMA cycle-steal timing diagram

3.6.3   Timer

Timer is also an important hardware and must be wisely utilized by ES designers for optimal operation and power savings. Timer hardware is driven by the clock of the MCU. However, there can be a pre-scalar that reduces the system clock speed prior to feeding to timer hardware. Timers can be set to count up or down. A common example of usage of timer is to trigger and interrupt after timer reaches terminal value (i.e., flips to 0 from maximum value). Timer is setup and started with timer control registers.

Some examples of timer control registers are Timer Interrupt Enable register (TIE), Timer Input capture/Output compare Select register (TIOS), Timer System Control Register 1 or 2 (TSCR1 or TSCR2), Timer Interrupt Flag 1 or 2 (TFLG1 or TFLG2), and Timer Count register (TCNT).

3.6.4   Watchdog Timer (WDT)

Watchdog timer (WDT) is a special purpose timer. By default, it is disabled. However, it can be enabled to detect a software crash. To use the watchdog timer, after enabling it, the ES designer must write code such that the watchdog timer is often refreshed/reset. The normally operating software will perform certain key actions within a fixed timeout interval. If the actions are not performed within the proper interval, the Watchdog forces the microcontroller into a reset sequence to restart the faulty software. This is important for ES as these systems often operate without any human intervention, thus must be able to detect software crash and restart the system by itself.

The ES designer must evaluate the duration of a "reasonable" timeout interval for the watchdog timer, and special instructions are inserted at key point in the program to "report in" to the Watchdog mechanism. This prevents the timer from reaching a count of 0 (because reaching 0 would reset the system) by writing to the period registers before the timeout occurs. This effectively reloads the WDT counter and start counting a new timeout interval.

3.7   Communications

Another important hardware aspect is to consider communications. Communications allow data and control signals to reach from one system to another. Communication can occur only through a medium. The medium can be guided medium (such as twisted pair wires, and optical fibers) or unguided medium (such as air, vacuum, and water).

Optical figure is a very high bandwidth communication channel, but relatively expensive and complex. Optical fibers can be multi-mode that is cheaper but has shorter range (such as campus deployments of internet) or single-mode that is more expensive but can reach long distance (such as intercontinental internet links). Copper wire-based communication (e.g., twisted pair wires) has good performance for shorter distance and very inexpensive. Co-axial cable has higher capacity but relatively bulky and expensive, whereas twisted pair is cheap but has lower capacity.

Wireless communication medium has the lowest capacity, very diverse, and the most difficult technology. However, it offers significant benefit especially for ES devices. Performance of a wireless link is determined by the physical world (e.g., objects, walls, buildings, etc.). Wireless link can reach very large distance, specially through lossless medium such as vacuum (e.g., inter-planetary satellite). Wireless technology can provide high speed in some case with limited range, but becomes expensive.

Some commonly used metrics for communication performance are listed below.

- **Half-power bandwidth (BW)** is the interval between frequencies at which $G_x(f)$ has dropped to half power (3 dB).

- *Equivalent rectangular bandwidth* (**W_n**) is defined by $W_n = P_x/G_x(f_c)$, where P_x is the total signal power over all frequencies.
- *Null to null bandwidth* is the frequency interval between first two nulls of $G_x(f)$.
- *Fractional power containment bandwidth* is the bandwidth with 0.5% of signal power above and below the band (FCC).
- *Bounded power spectral density* is the band defined so that everywhere outside $G_x(f)$ must have fallen to a given level.
- *Absolute bandwidth* is the interval that contains all of the signal's frequencies.
- *Signal-to-noise ratio (SNR)* is the ratio of average signal power to average noise power and represented as $SNR(dB) = 10 \log_{10} P_{signal}/P_{noise}$.
- *Shannon Channel Capacity* is defined by $C = BW \log_2 (1 + SNR)$ and represented in bits per second.

3.7.1 Wired Communication Topologies

In previous Sects. 3.5.4 and 3.5.5, we have discussed various wired topologies. Among wired communication, serial and parallel communications both are used in ES. Common wired communication topologies include UART (or USART), I2C, USB, SCI, SPI, Ethernet, etc.

3.7.2 Wireless Communication Topologies

For newer ES devices, wireless communication techniques are becoming more important as these devices are becoming part of IoT, wearables, smart devices, and sensornet. Wireless technologies are growing in applications. These technologies differ in terms of frequency of operation, communication range, data throughput, and power consumption. Among various wireless communication topologies, RF point-to-point link is not suitable for most ES devices. Some of the more suitable topologies are mentioned below.

3.7.2.1 Cellular Data Network (SIM)

This topology uses cellular network or cell-towers, typically used for mobile phones. This requires use of SIM card or equivalent technology and requires subscription to a cellular provider (recurring expense during runtime for subscription). Related standard is IEEE 802.16 and IEEE 802.16a. Example technologies in this topology are 3G (e.g., EDGE), 4G (e.g., LTE, WiMax), and 5G (upcoming). This topology allows the wireless ES device to connect to internet. As cell service is available in vast area of the country, this technology allows access to internet for ES systems that cannot be limited to predefined geographical areas, such as autonomous cars.

3.7.2.2 WiFi

This technology also allows internet access (through access point or routers), but does not have recurring subscription cost. This follows IEEE 802.11 standards. WiFi spectrum is divided in 13 + 1 channels. As these channels overlap in spectrum, it is desirable to find the mostly free spectrum for less interruptions. Most IoT uses this technique to connect to internet. Each WiFi module will have a MAC address (unique hardware identification number) and an IP address (unique internet identification number). Older IP standard is IPv4 (32-bit address), which is being replaced with new standard IPv6 (128-bit address with backward compatibility). WiFi can operate in 2.4 GHz frequency (lower throughput, larger range) or 5 GHz frequency (higher throughput, smaller range). An example of this type of ES device is smart thermostats. The ES designer must have provision such that user can store the WiFi network name and password to the ES device after deployment.

3.7.2.3 Bluetooth (BT)

This technology works in very short range and for low data throughput. This is suitable for device to device communication, such as a user's wristband to communicate with the user's smartphone. The related standard is IEEE 802.2 and it follows OSI layer equivalent. BT uses 2.4 GHz frequency, but has different channels to WiFi so that there is no interference. There are mainly 3 classes of BT topologies: Class 1 for longer distance (up to a km), Class 2 for shorter distance (up to 100 m), and Bluetooth Low Energy (BLE) for short distance low-power communication. Almost all smartphone uses BT Class 2 modules. BT communication requires selection of a profile by the ES designer that matches with the receiver profile. An example profile for beginners is SPP (Serial Port Profile) which can be treated as a simple serial port.

3.7.2.4 Sensor Nodes

Some dedicated topologies were developed for sensor nodes with specific focus on mesh networking and low-power operation. Examples of these topologies are ZigBee or XBee, and Mote. ZigBee is a standard that defines a set of communication protocols for low-data-rate, very low power, short-range wireless networking. Its standard is IEEE 802.15.4b and highly suitable for mesh networking. ZigBee nodes can be PAN coordinator, full function device, and reduced function device. It can form various wireless networks such as Star, Mesh, and Cluster-tree. Sensornet or Mote are also suitable for distributed sensor networks. Another topology was dedicated for low-power sensors and wearable devices called Ant+. However, recently WiFi and BT modules are dominating these applications as well, thus these dedicated topologies have very limited usage.

3.7.2.5   RFID or NFC

Radio Frequency Identification (RFID) technology was developed for smart ID card. RFID can be three types: Active, Passive, and Semi-active (or Semi-passive). Active has the largest range (up to 10 m) but requires battery in the sensor. Passive type does not have any battery in the sensor, but has very short range of operation (less than 1 m). The semi-active (or semi-passive) has battery, but only activates when a request is received, thus maximizing battery life. Most RFID we commonly see are passive types. Near Field Communication (NFC) is another similar technology which has found applications for smart wallet.

3.7.2.6   GPS

Global Positioning System (GPS) is another sensor system; however, this is only one-way communication from satellite to GPS sensor and has very specific application of location determination. GPS sensors can approximately locate its position within typically a few meters range. One limitation of GPS is that it requires line of sight to satellites, thus does not operate indoors.

3.7.2.7   Performance Metrics

Some performance metrics used in wireless communication are latency (inherent delay or response time), and throughput. Latency indicates the time between start and end of a task or operation. Throughput denotes tasks per second. Other metrics are clock frequency, instruction per second, range, and power consumption.

3.7.2.8   Wireless Security

All wireless technologies suffer from security issues, as any intruder within the range can receive the signal and use it maliciously. Some techniques are incorporated in various topologies to reduce this type of cyber-security issues, such as authentication requirement for access control, confidentiality, and integrity from protocol spoofing. Examples of authentication rules are WEP, IPSEC, SSL, EAP, etc. Furthermore, algorithms can run on a process with encryption, hash, or with unique key (e.g., RSA). Key is typically a string of numbers. Another technique used in BT is frequency hopping in a predefined sequence which would be unknown to an intruder. Furthermore, data encryption can be performed such as plaintext is converted to a ciphertext using encryption algorithm that requires specific decryption algorithm to decode. This is commonly done with public key and private key. Public key is known to everyone, whereas private key is only known to the user. As computing devices are growing in complexity, multilevel security and/or multifactor authentication are becoming more effective. However, ES designers mostly rely on standard security algorithms already in place, rather than designing their own security algorithms. Cyber security is a demanding area and readers can explore related textbook further for details of these algorithms.

There are some other communication topologies such as light-based or sound based. However, these have very limited applications in ES, thus not elaborated here.

3.8   Pulse Width Modulation (PWM)

Some MCU contains PWM hardware can be utilized for a number of purposes, such as generating various rates of blinking of LEDs by changing frequencies, analog output voltages by filtering the PWM output, and controlling DC motor speed of a RC car by changing the duty cycles of PWM waves. Independent PWM channels with programmable period and duty cycles are available in a lot of MCU. Programmable PWM can be enabled or disabled for each channel. Software selection of PWM can also specify duty pulse polarity for each channel. Period and duty cycle are double buffered, thus change takes effect when the end of the effective period is reached, i.e., PWM counter reaches to zero or when the channel is disabled. Also, programmable center or left-aligned outputs on individual channels can be set. Different bit-width determines PWM resolution or duty cycle ranges. Various MCUs provide a wide range of frequencies for PWM with programmable clock select logic.

3.9   Battery Considerations

Some ES devices needs to be operated with battery, for example wearables. This section list some of the related considerations that ES designers must be aware of.

3.9.1   Battery Technologies

Within various battery technology (battery chemistry), currently Lithium Polymer (LiPoly or LiPo) is the most suitable for ES powering with recharging needs. This type of battery produces a nominal 3.7 volts, and should not be combined (neither series nor parallel). LiPo batteries have high energy density, and low charging time.

For LiPo rechargeable battery, ES designers need to incorporate a power management chip, as LiPo batteries must not be overcharged or fully discharged. For example, MCP73831T LiPo battery charger is a simple to use power management chip.

For non-rechargeable batteries, carbon-based batteries (such as AA, AAA, C, D) produces 1.5 V per cell, and coin cell battery (e.g., CR2032) produces 3 V. These batteries are okay to be combined in series or parallel.

3.9.2   Power Saving Modes for MCU

For most ES devices, especially for battery powered devices, it is extremely important to optimize power that will directly affect hours of operation. Other factors related to power optimization is reduced thermal output, improved usability, and higher access to the system. To save power consumption, it is recommended to operate an MCU at the lowest clock frequency that can guarantee timing requirements, as higher clock frequency will lead to higher power consumption. Another consideration is reduced supply voltage. For example, reducing supply voltage from 5 to 3.3 V save ~56% power. However, ES designers need to ensure all electronic components will operate with the reduced supply voltage.

Another important power optimization technique for MCU is utilizing "Sleep Modes." A uP should be kept in highest level of sleep mode as much as possible to save energy. Common operating modes of uP are Active mode, Idle mode, Noise reduction mode, and Power-down (sleep) mode. For instance, ATtiny88 consumes 0.2/1.4/4.5 mA for 1 Mhz, 2 V/4 MHz, 3 V/8 MHz, 5 V operation in Active mode, respectively. With Idle mode, the current consumption reduces to 0.03/0.25/1 mA, respectively. For power-down mode, the current consumption can be 4 uA with WDT enabled, and < 0.2 uA with WDT disabled. Note that some peripherals are shut down in some of these modes. ES designer must consult the MCU datasheet for details.

3.10   SPICE Simulation and PCB Design

This section discusses a typical custom Printed Circuit Board (PCB) design procedure for hardware of embedded system. A free PCB design software, KiCad, is used for examples (version 5.0.2). See Appendix A for a quick tutorial of KiCad. In this first example here, a simple op-amp based amplifier schematic design and simulation. Later example will show a bandpass schematic filter design, simulation, and PCB design.

In KiCad, the tools that we will be using are Eeschema (for schematic design) and Pcbnew (for PCB design). Figure 3.21 shows the relation of various activities that will be done with each of these tools.

Fig. 3.21 Procedure for using Eeschema and Pcbnew KiCad tools

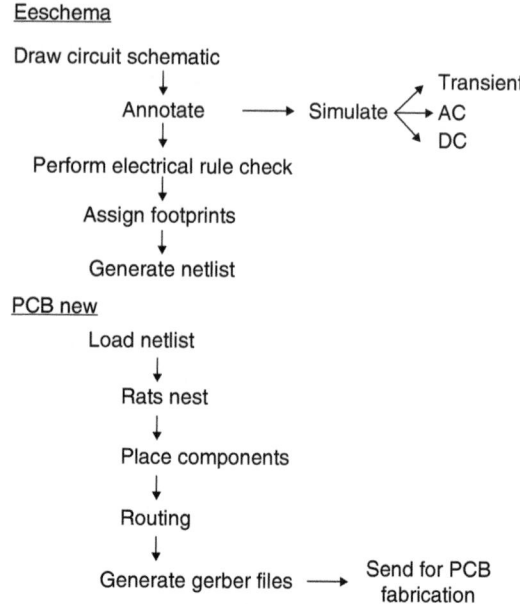

Eeschema

Draw circuit schematic

Annotate ⟶ Simulate

Transient
AC
DC

Perform electrical rule check

Assign footprints

Generate netlist

PCB new

Load netlist

Rats nest

Place components

Routing

Generate gerber files ⟶ Send for PCB fabrication

3.10.1 Schematic Design

Design the following schematic with KiCad Eeschema (schematic designer) software (Fig. 3.22). Here, two power supplies are used (2.5 V DC each) for the positive and the negative supplies of the op-amp. As the input, a sine wave of 1 kHz signal, 10 mV amplitude, 0 V offset, is provided. The ratio of R2 and R1 sets the gain of -10. A 1 kΩ load resistance is used at the output. Use the "0" symbol from Pspice library for ground terminals.

Use the opamp symbol from Pspice library of KiCad. It is a 5-terminal opamp design. For the opamp model, use the provided ideal opamp model (provided in Fig. 3.23) saved in a text file named "idealopamp.cir." Note the power supply pins are not used at all in this ideal model! Associate the model with the opamp in the schematic diagram (Properties > Edit Properties > Edit Spice Model > Model > Select file to load the model file).

3.10.2 Schematic Simulation

From KiCad, annotate the design, then simulate it. To simulate a schematic design with time as x-axis (called "Transient simulation"), first set your AC signal source in schematic with the required signal amplitude and frequency (Properties > Edit Properties > Edit SPICE Model > Source tab > Sinusoidal). Then open the

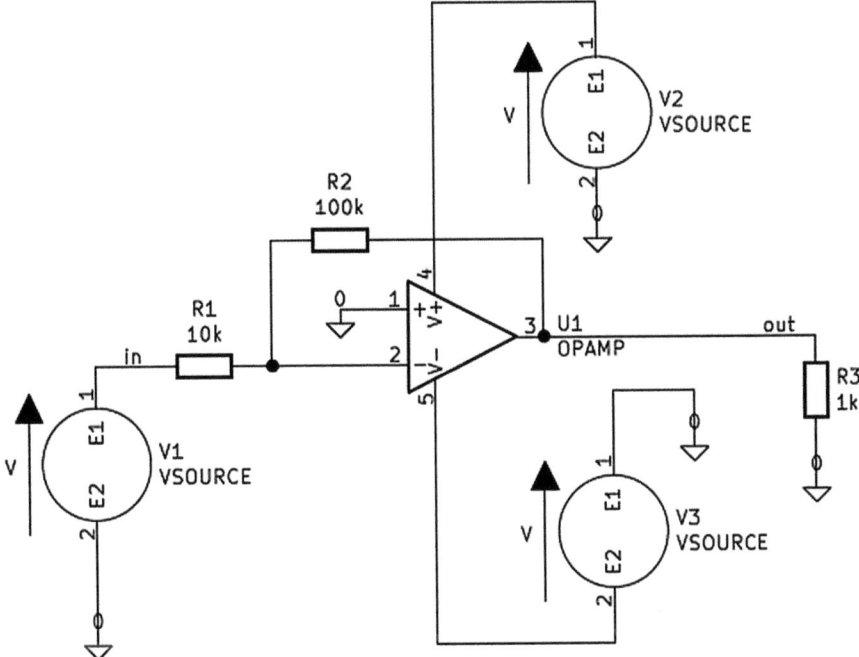

Fig. 3.22 Schematic diagram of the simple inverting amplifier

Simulator tool (Tools > Simulator), go to Settings of the simulator tool, then click "Transient." You can specify the start time (default 0), end time (e.g., 10 times of signal period), and step size (typically 1000 times smaller than end time). After this, run the simulator, then click on Signals to pick the signals of your choice (e.g., Vin, Vout). You should be able to see the transient simulation.

Figure below shows transient simulation output (with settings of 10 u for Time Step, and 10 m for Final time). This shows that the output is 10 times higher than the input, and the phase is inverted, as expected (Fig. 3.24).

If you want to see frequency plot (aka Bode plot), you will need to perform an AC simulation. For this setup, change the AC source by removing sinusoidal settings, and put DC = 0 and AC = 1 in DC/AC Analysis (Properties > Edit Properties > Edit SPICE Model > Source tab > DC/AC analysis). After this, start the Simulator tool. In the Settings, click AC tab, then set the start frequency and end frequency as you need (e.g., 1 Hz to 10 k Hz). Also enter number of points (typically 1000 points provide a smooth curve). Click simulation button, then add the signal you want to observe response for (e.g., Vout).

```
*
* OPAMP MACRO MODEL, SINGLE-POLE
* connections:         non-inverting input
*                      |    inverting input
*                      |    |    output
*                      |    |    |  (supplies)
.SUBCKT OPAMP1         1    2    6   101   102
* INPUT IMPEDANCE
RIN            1        2        10MEG
* DC GAIN (100K) AND POLE 1 (10HZ)
EGAIN          3        0        1        2        100K
RP1            3        4        1K
CP1            4        0        15.915UF
* OUTPUT BUFFER AND RESISTANCE
EBUFFER        5        0        4        0        1
ROUT           5        6        10
.ENDS
```

Fig. 3.23 Model file for an ideal opamp

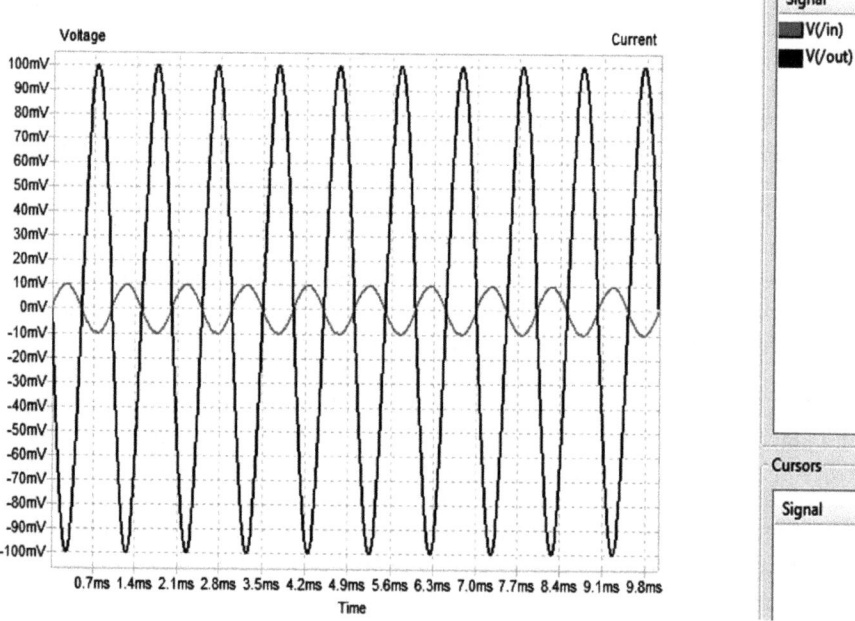

Fig. 3.24 Transient simulation output of the simple inverting amplifier

3.10.3   PCB Design Based on a Schematic

Printed Circuit Board (PCB) is a planar mechanical board with electrically conductive (copper) traces for creating complete electronic circuits by soldering the electronic components. PCBs can have single, double or multiple layers. Single layer PCBs have copper layer on one side and are suitable for newbie/hobby projects. Double layer PCBs are suitable for low frequency circuits and most widely used. In this type of PCBs, copper layers exist in both top and bottom side of the substrate. Another type is 4-layer PCBs that are suitable for high frequency circuits or mixed signal circuits. This are relatively expensive and only be used if needed.

A typical PCB has the following layers:

- Substrate (typically flame retardant, FR4).
- One or more layers of copper, which is etched in accordance to the PCB design (Gerber file) by the hardware designer.
- A solder mask layer that protects the copper traces where solder will not be applied.
- A silk screen layer that can be used to write notations, symbols, and labels.
- Drill, typically for via that connects different layers of copper.

In this example, we will create a 2-layer PCB for an inverting amplifier (analog components only). First, create the schematic shown in Fig. 3.25. This is an inverting amplifier design using an op-amp active low pass filter with a cut off frequency $f_c = \frac{1}{2\pi(100k)(1n)} = 1.59\ kHz$. The inverting amplifier gain is set to $A_v = -\frac{100k}{1k} = -100$.

The two capacitors on the power supply line are included to reduce supply noise (e.g., ripple) with values of 1 μF and 1 nF. The 1 μF capacitor will be an electrolyte type, while the 1 nF capacitor will be a ceramic type.

A 10 V unipolar supply voltage is used with a virtual ground generator circuit that has a resistive voltage divider (of 100 kΩ each) followed by an op-amp based buffer circuit. The output of this buffer is at 5 V and will be used as 0 node (connected to ground symbol of the schematic).

Set the signal source with 1 kHz sine wave with 0 offset and 10 mV amplitude. A 1 kΩ load resistance is used at the output of the op-amp for simulation only. The input and output nodes are labeled as *"vin"* and *"vout,"* while the power supply nodes are labeled as *"vdd"* and *"vss."*

For both op-amps, ensure that a SPICE model is set in their Properties with correct pin configuration. An example of a modified SPICE model based on TI LM258 chip is provided in Fig. 3.26. This is a more practical op-amp model (compared to ideal op-amp in the first example) that will more closely represent the actual behavior of the circuit.

For simulation, open the KiCad SPICE simulator (under Tools menu). Use Transient simulation with Time step: 10 μs, and Final time: 10 ms. Run the simulation, and add *"vin"* and *"vout"* to the waveform window. It should look like Fig. 3.27.

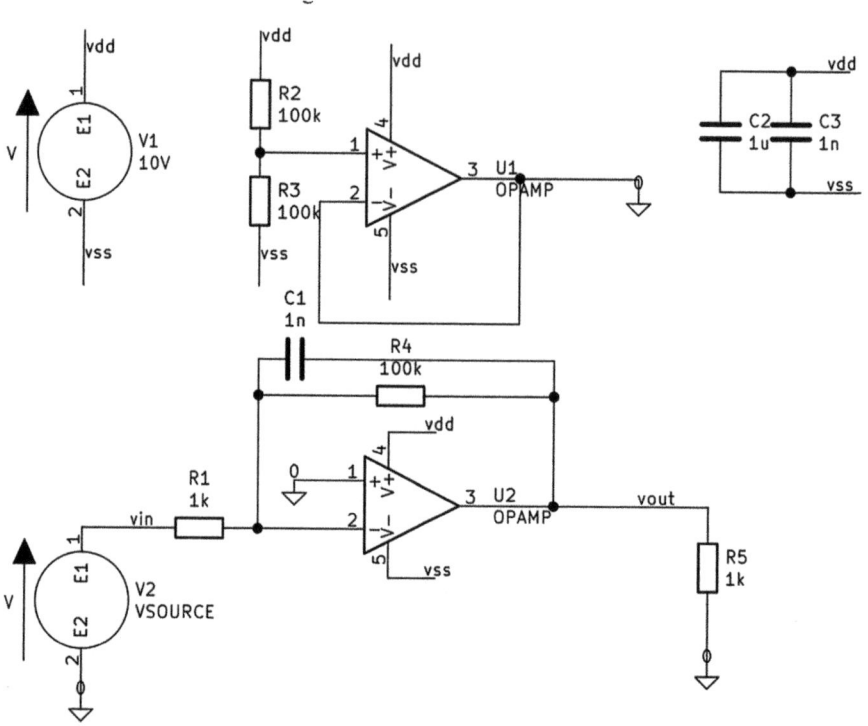

Fig. 3.25 Schematic diagram of the inverting amplifier for the analog PCB design example

Once the simulation is satisfactory, and you are absolutely complete with schematic, it is time to start the PCB design. For this, we will need to slightly modify the schematic. You can copy the schematic folder, and work on the copied folder to preserve your simulation-able schematic intact.

We will remove the voltage supply symbol, and replace it with a Barrel Jack connector symbol. We will also remove the signal source, and replace with a 2 pin header, and remove the load resistance (R5) and also replace with a 2 pin header. Other than that, we will use LM358 dual op-amp chip. In the corresponding KiCad symbol, you will notice that it has 3 parts: Part A, Part B, and Part C. Parts A and B are the two op-amp of the chip, and the Part C is the power supply connected. Insert all of these three parts, and replace the two PSPICE op-amp with Part A and B, while connect *vdd* and *vss* wire to the Part C, as shown in Fig. 3.28. Annotate the new schematic. When you simulate Electric Rule Check (ERC), two arrows will appear to Part C, as they are not connected to any component as seen in Fig. 3.28. You can ignore this warning.

Next, we will assign footprints to the components. Click CvPCB icon in the KiCad menu. From the list of footprints shown on the right, we will need to select the appropriate ones for the actual components that we will use. Consulting the datasheet

```
.SUBCKT LM258_ON    1 2 24   11   12

***** Input Stage *****
Q_Q1   4 5 6    QPNP1
Q_Q2   7 8 9    QPNP2
I_I1   111 10  DC 1m
R_RC1  4 12    95.49
R_RC2  7 12    95.49
R_RE1  10 6    43.79
R_RE2  10 9    43.79
V_Vio  2 8     0Vdc
E_E4   1 5 16 14       1
E_EVcc 111 0 11 0      1
G_Vcc  0 11    poly(1) 11 0    0.556m 4.8u

***** Gain Stage & Frequency Response Stage *****
R_R3   13 12   1k
R_R4   111 13  100k
E_Eref 14 0 13 0       1
G_G1   14 15 7 4       0.01047
R_Rc   14 15   9.549Meg
C_Cc   14 15   1.667n

***** Output Stage *****
E_E1   22 14 15 14     1
V_F1   23 24   0
F_F1   11 0 V_F1       1
R_Rout 22 23   13

***** Common Mode Rejection *****
R_R1   3 1     1Meg
R_R2   2 3     1Meg
G_G2   14 16 3 14      5.6234n
R_Rcmr 17 16   10k
L_Lcmr 14 17   1.59H

***** Output Voltage Limiting *****
D_D1   15 18   D10D1
D_D2   19 15   D10D1
V_Voh  111 18  2.183
V_Vol  19 12   0.639

***** Output Current Limiting *****
D_D3   15 21   D10D1
D_D4   20 15   D10D1
V_V3   21 23   -0.207
V_V4   23 20   -0.467

.model QPNP1   PNP(Bf=10841.7)
.model QPNP2   PNP(Bf=11741.7)
.MODEL D10D1 D IS=6E-16 RS=1.000E-3 VJ=.75 BV=100E6
.ENDS
```

Fig. 3.26 An example op-amp SPICE model (modified from LM258 model provided by TI)

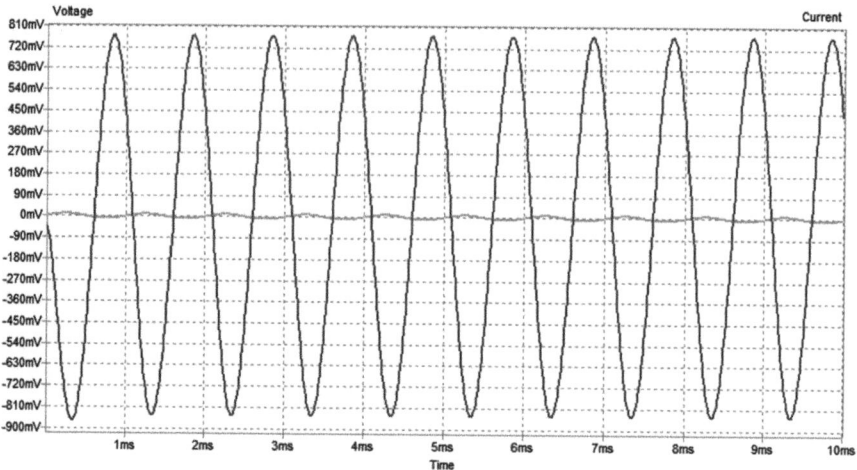

Fig. 3.27 Transient simulation waveform from KiCad

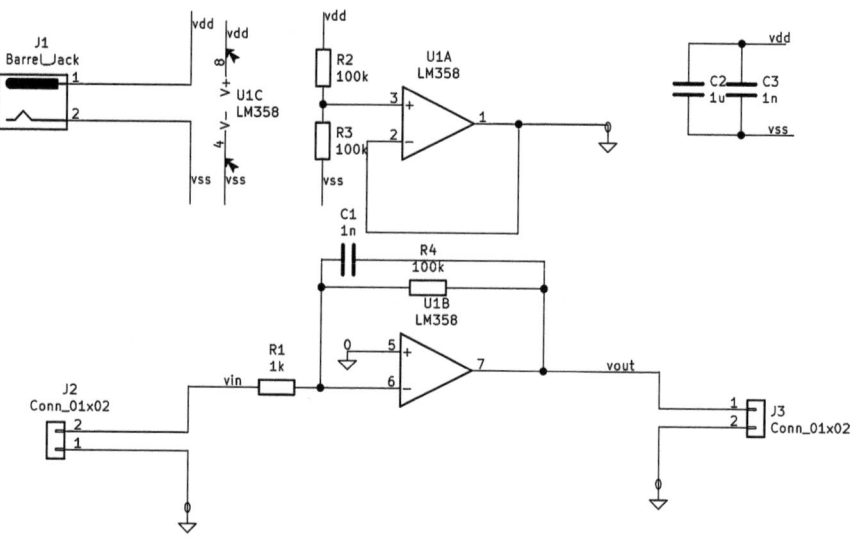

Fig. 3.28 Modified schematic for PCB design

of the components, determine the footprints. In this example, we will use the footprints as shown in Fig. 3.29 that are thorough hole type (THT). All components must have assigned footprint before proceeding further. Note that if the footprint that you require is not available in the KiCad library, you might need to prepare custom footprint with proper pad-stack. Refer to KiCad manual as this is beyond the scope of this discussion here.

```
Symbol : Footprint Assignments
   1        C1 -                 1n : Capacitor_THT:C_Disc_D3.0mm_W2.0mm_P2.50mm
   2        C2 -                 1u : Capacitor_THT:CP_Radial_D5.0mm_P2.00mm
   3        C3 -                 1n : Capacitor_THT:C_Disc_D3.0mm_W2.0mm_P2.50mm
   4        J1 -        Barrel_Jack : Connector_BarrelJack:BarrelJack_CUI_PJ-063AH_Horizontal_CircularHoles
   5        J2 -         Conn_01x02 : Connector_PinHeader_2.54mm:PinHeader_1x02_P2.54mm_Vertical
   6        J3 -         Conn_01x02 : Connector_PinHeader_2.54mm:PinHeader_1x02_P2.54mm_Vertical
   7        R1 -                 1k : Resistor_THT:R_Axial_DIN0204_L3.6mm_D1.6mm_P7.62mm_Horizontal
   8        R2 -               100k : Resistor_THT:R_Axial_DIN0204_L3.6mm_D1.6mm_P7.62mm_Horizontal
   9        R3 -               100k : Resistor_THT:R_Axial_DIN0204_L3.6mm_D1.6mm_P7.62mm_Horizontal
  10        R4 -               100k : Resistor_THT:R_Axial_DIN0204_L3.6mm_D1.6mm_P7.62mm_Horizontal
  11        U1 -              LM358 : Package_DIP:DIP-8_W10.16mm
```

Fig. 3.29 Footprints associated with each components of the schematic in KiCad CvPCB interface

After footprint is assigned, create a netlist for the model using NET icon (or Tools-> Generate Netlist File). Ensure the tab says "PCBnew," notice the filename (or change it if you want), then click Save. A netlist appropriate for PCB design will be exported (by default, it will be in the project folder).

You also generate the Bill of Materials (BOM) at this point by clicking the BOM button (or Tools-> Generate Bill of Materials). You will need to include a Plugin for this. You can use the default plugin from KiCad, which can be found in Kicad installation folder (../KiCad/Bin/scripting/plugins/bom_csv_grouped_by_value.py). By default, the generated file will not have any extension. To save the file extension, in the command line statement, add .csv to %O (i.e., "%O.csv" will be the last entry of the command line). When you click Generate, a csv file will be saved that will show the components that you need, similar to Fig. 3.30.

You are done with the schematic portion of the PCB design process, so close it, and open the PCBnew software (now called PCB Layout Editor). You can set the page to US letter for printing if you need (File- > Page Settings - > US Letter).

First, we will enter the design rules and create standard tracks that we will use during the PCB design. To do this, go to Setup- > Design Rule. Here, you need to enter the proper values based on your choice of PCB foundry design rules. In this example, we will use OSHPark design rules. The minimum trace width and minimum distance between traces are mentioned as 6 mils. As KiCad uses mm measurement, we will enter 0.15 for Clearance and 0.15 for Track width. We can keep the rest untouched, as drill dimensions of OSHPark is within the default values of KiCad.

Go to Global tab and also change the Minimum track to 0.15. Also create some Track widths which will be handy during the design process (for example Track 1: 0.15, Track 2: 0.2, Track 3: 0.25, Track 4: 0.3, etc.).

Now we are ready to import the netlist that we have saved. Click NET icon (or Tools-> Load Netlist), and select the netlist file that you exported before. Click Read Current Netlist, and then place the imported netlist components to any place within the PCB design page. When you click Zoom Fit, you will be able to make out the components. It should look something like Fig. 3.31. Note that the components are too close to each other for soldering, as well as the connections are symbolic

Collated Components:						
Item	Qty	Reference(s)	Value	LibPart	Footprint	Datasheet
1	2	C1, C3	1n	Device:C	Capacitor_TH ~	
2	1	C2	1u	Device:C	Capacitor_TH ~	
3	1	J1	Barrel_Jack	Connector:Barrel_Jack	Connector_B ~	
4	2	J2, J3	Conn_01x02	Connector_Generic:Conn_01x02	Connector_P ~	
5	1	R1	1k	Device:R	Resistor_TH1 ~	
6	3	R2, R3, R4	100k	Device:R	Resistor_TH1 ~	
7	1	U1	LM358	Amplifier_Operational:LM358	Package_DIP	http://www.ti.com/lit/ds/symlink/lm2904-n.pdf

Fig. 3.30 The BOM file will have the list of components in collated form, ready for parts ordering

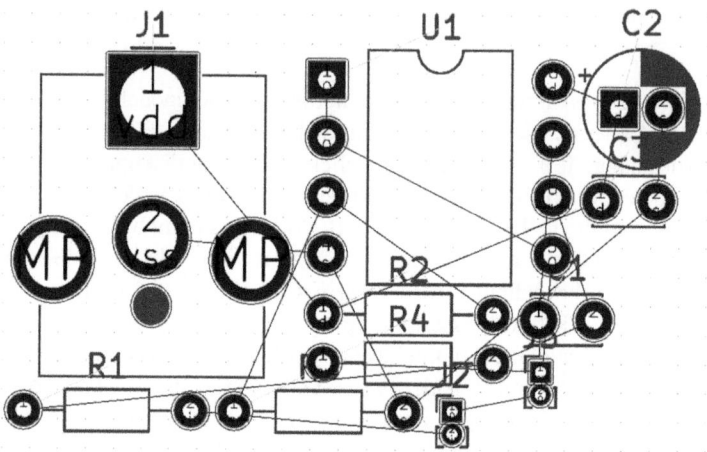

Fig. 3.31 Imported netlist in the form of Rats Nest in the PCB Layout interface

(white lines). This representation of the components in the PCB layout is called "Rats Nest."

In preparation for "Placement" of the components properly in our PCB design, first we will create a border of the PCB. For this, start with selecting Edge.cuts from the top menu bar drop down button, then click on Graphic Line button on the toolbar to the right, then draw a rectangular area about 30–40% larger than the area seen on the Rats Nest. You can change the edge later, so start with an approximate size that you expect your PCB will end up to.

After you created an area for PCB, next is to drag each of the components on your PCB area and position them as you think appropriate. Typically, all terminals, sockets, and connectors are placed near the edge, and leave a few mm spacing between parts to allow for soldering process. In this example, we will place all components on the top side, although placing components on both sides will make the design compact, but complicated which you can try later as you become expert.

A picture of the layout is shown in Fig. 3.32. In this figure, the symbolic connections are not seen due to resolution/color conversion issue, but in your layout you should be able to still see lines connecting pins that represents how they need to be connected (based on your schematic design). Also note that you can move and

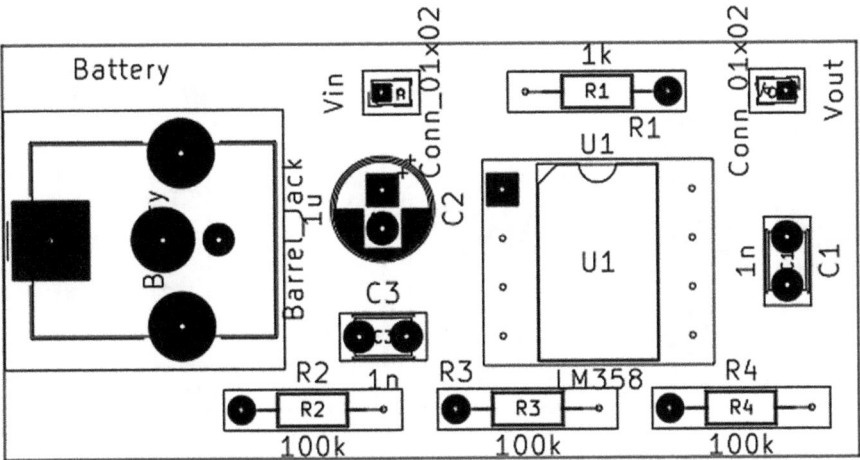

Fig. 3.32 PCB layout after placement

edit the silk screen texts (e.g., here J1 is changed to Battery and moved inside the PCB area, J2 is changed to Vin, and J3 is changed to Vout). Some of the text in this view are comment (e.g., Conn_01x02) and they will not show up in actual PCB, so do not worry. You can also put your own text by selecting the silkscreen layer (e.g., F.SilkS) from the top menu drop down box and selecting "T" icon for entering text.

Once you are happy the placement of the footprints, we will start drawing the traces (except ground, or 0 in this example). To create a trace, select F.Cu (Front copper) from the top drop down menu, then select the green line icon (Route traces) from the right side menu. Click on the center of a pin and route the traces to the other pin as shown by the symbolic white line. You will notice that the symbolic white line will disappear as you draw the trace. Avoid using 90° bend, try to use only 45° bends. Use 0.2 mm trace width for all lines except vdd and vss, for which you can use 0.3 mm (for higher currents). To erase any trace, click to select the trace segment and use Delete key to remove the trace segment. You can route your trace under the IC and resistors as long as they do not connect the pin terminals. As you complete connections, you will notice that some of the traces might require you to use the bottom layer of the PCB. To use the bottom layer while routing (to avoid shorts), press "v" then click where you want to place a via (that will connect the top layer to the bottom layer copper). The trace color will automatically change to B.Cu (Bottom Copper). When you want the trace to come back to top layer, repeat the process of via placement (i.e., press "v" then click where you to place it). Note that it is better to minimize the number of vias when laying out traces. Continue this process until you eliminate all symbolic wires except 0 node wires. Your PCB design should look like Fig. 3.33.

We will use ground fill for 0 node. This will cover all open space will copper layers and connect them to 0 node. This improve noise response and minimizes wastage of materials during the fabrication process. To place the ground fill (of top

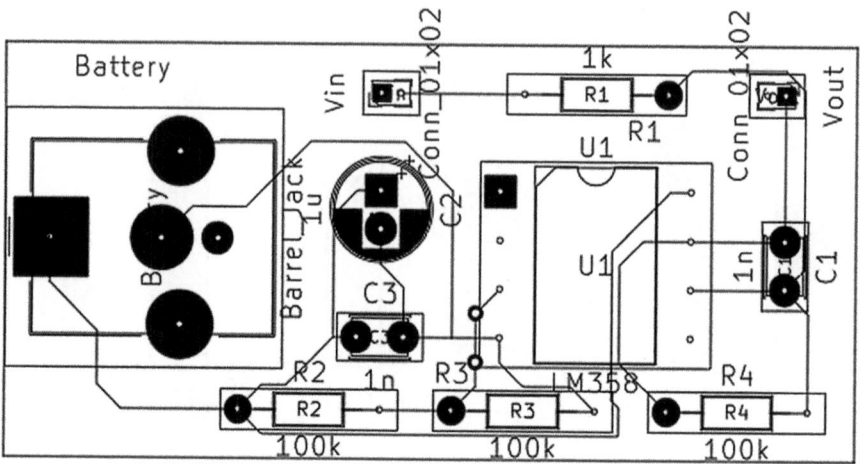

Fig. 3.33 PCB layout after placement traces are routing (except 0 node)

copper layer), select F.Cu as the layer in the top menu drop down box, click Add Zones button on the right side toolbar, then click the top right hand corner of the PCB. This will open a dialog box. Select F.Cu and 0 node. Also select Default Pad: Thermal relief, and Outline slope: H, V, 45° only. Then click OK. Continue clicking the remaining three edges of the PCB in sequence. Once you finish clicking the top left corner at the end, the filled layer will show up. Note that there might be still symbolic white line remaining on the PCB, that you will need to manually eliminate by routing wires from one of the pins from one side of the filled shape to the other side of the filled shape. You will need to use the bottom layer (using via- > bottom layer- > via) for these routings. Once you complete your PCB design, it should look like Fig. 3.34.

You can check different net connections by selecting the Highlight Net icon for the right side menu, and clicking on any trace. You can also hide layers by unchecking the corresponding layer on the right side Layer Manager toolbar. Once you ensure all connections are complete, you can also check for design errors by clicking Perform Design Rule Check icon (or Inspect- > Design Rule Checker). The design rule check (DRC) will show if there is any error. If no error message shows up, your design is complete, otherwise you will need to fix the errors.

You can view the 3D representation of you potential PCB using the 3D viewer (View- > 3D Viewer). The top and bottom side should look like Fig. 3.35. Note that not all footprint library contains 3D view entry, hence some components will not show up (e.g., the Barrel Jack in this example).

The final step is to create the gerber files (artwork) that fabrication facility utilize to make the PCB. For the PCB, only the following layers will be utilized: F.Cu, B. Cu, F.Mask, B.Mask, F.SilkS, B.SilkS, Edge.cuts. In addition, we will need to generate Drill layout.

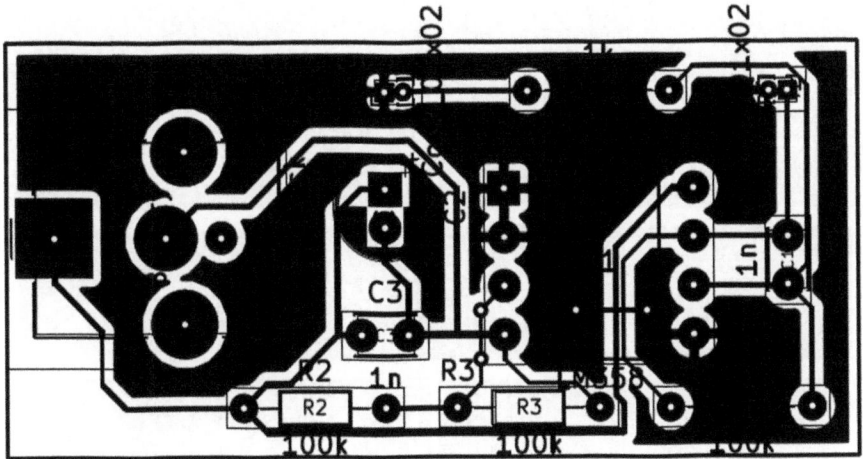

Fig. 3.34 PCB layout after completion of routing

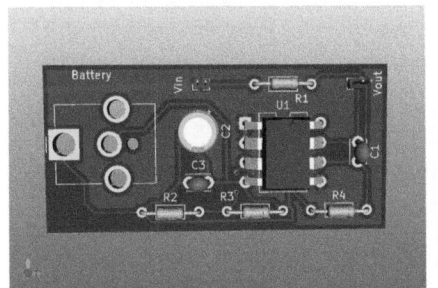 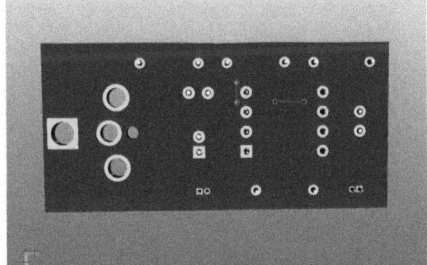

Fig. 3.35 3D view of the PCB

To prepare gerber file, select File- > Plot. The default plot format is gerber. For output directory, type in "gerber" so that the generated files will be placed in a folder named "gerber" inside your project directory. Make sure the needed layers are selected (as mentioned above). Click Plot. To generate the drill file, click Generate Drill File. Then click on Generate Drill File in the new dialog box. Close both dialog boxes.

To submit gerber file to the fab, go to the gerber folder, and select all files to create a single zip file. Log in to PCB fabrication website (e.g., OshPark.com), and upload this zip file. Check to see if any error is shown. OshPark shows the different layers that is generated from the uploaded gerber files, as shown in Fig. 3.36.

Check and compare with PCB layout outputs in OshPark website and compare them with KiCad layout and 3D viewer to ensure the uploaded version is correct.

You can also use surface mount device (SMD) components that are more optimal for space utilization. Below is another example of PCB design for the given schematic with SMD components (Schematic and footprint are shown in Fig. 3.37 and PCB layout and 3D view are provided in Fig. 3.38).

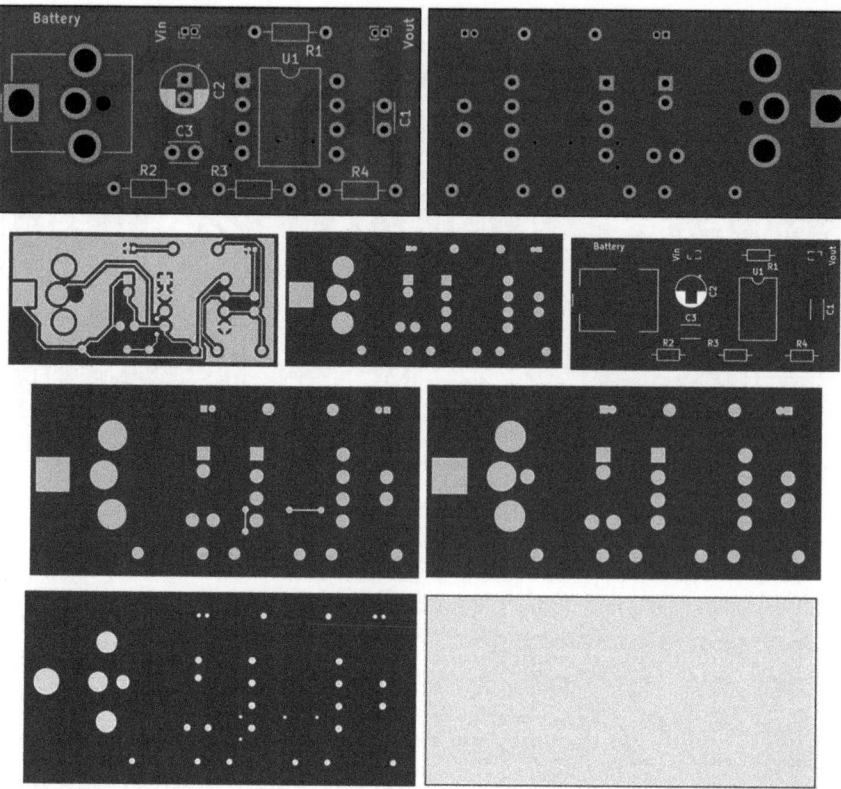

Fig. 3.36 OshPark PCB views (from top left to right: Board Top, Board Bottom, Top (copper) layer, Top Solder Mask, Top Silk Screen, Bottom (copper) Layer, Bottom Solder Mask, Drills, Board Outline). (Note that this design does not contain a bottom silk screen layer)

3.11 Examples of a Practical AFE

This chapter provided a brief description of hardware aspects in relation to embedded system AFE. Some information is also provided for back-end actuators. Furthermore, some simple design examples were provided. This by-no-means provides complete information required for ES designers, however, just serves as guidance to organize the design process and some important factors to consider. For complete design process, ES designers need to consult related engineering reference materials, software, and design process.

As an example of AFE for practical device, below is the AFE design of an electroencephalography (EEG) data collection system (Fig. 3.39) [3]. This can be modified with different gain and filter cutoff range to collect electrocardiography (ECG/EKG) or electromyography (EMG) signals as well.

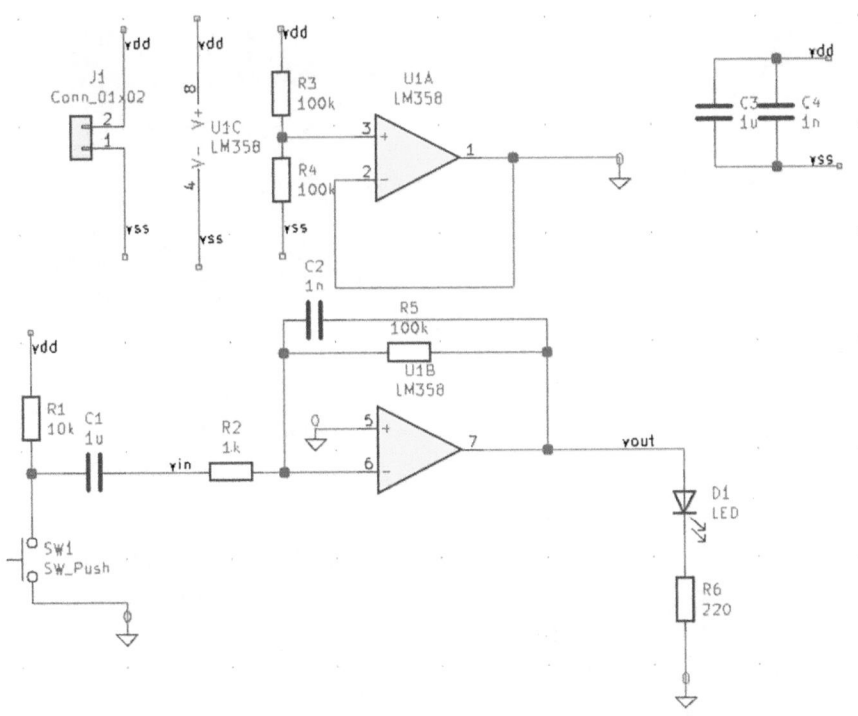

Symbol : Footprint Assignments			
1	C1 –	1u :	Capacitor_SMD:C_0603_1608Metric
2	C2 –	1n :	Capacitor_SMD:C_0603_1608Metric
3	C3 –	1u :	Capacitor_SMD:C_0603_1608Metric
4	C4 –	1n :	Capacitor_SMD:C_0603_1608Metric
5	D1 –	LED :	LED_SMD:LED_0805_2012Metric
6	J1 –	Conn_01x02 :	Battery:BatteryHolder_Keystone_103_1x20mm
7	R1 –	10k :	Resistor_SMD:R_0603_1608Metric
8	R2 –	1k :	Resistor_SMD:R_0603_1608Metric
9	R3 –	100k :	Resistor_SMD:R_0603_1608Metric
10	R4 –	100k :	Resistor_SMD:R_0603_1608Metric
11	R5 –	100k :	Resistor_SMD:R_0603_1608Metric
12	R6 –	220 :	Resistor_SMD:R_0603_1608Metric
13	SW1 –	SW_Push :	Button_Switch_SMD:SW_MEC_5GSH9
14	U1 –	LM358 :	Package_SO:SO-8_5.3x6.2mm_P1.27mm

Fig. 3.37 Schematic and footprint of a modified circuit with SMD components for PCB design

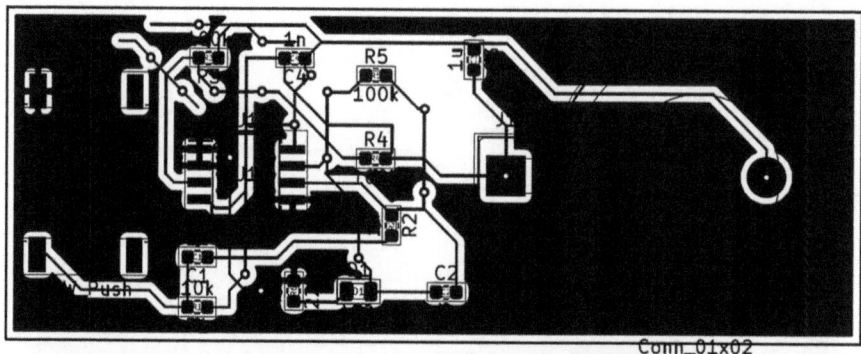

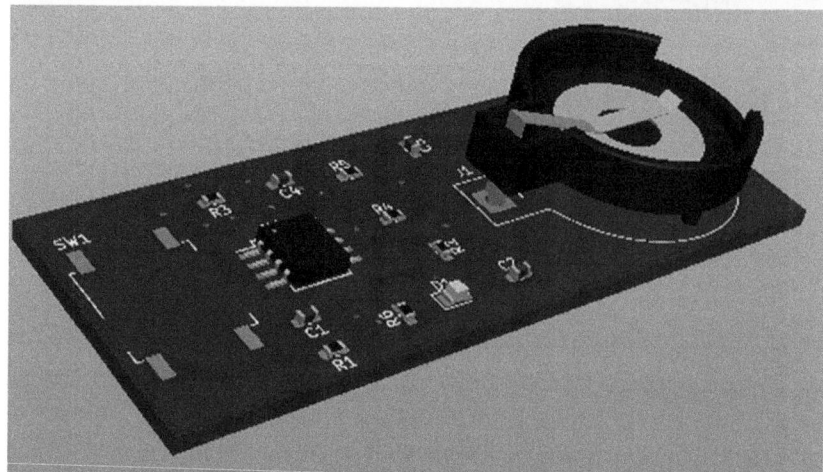

Fig. 3.38 PCB design and 3D view of the modified circuit with SMD components

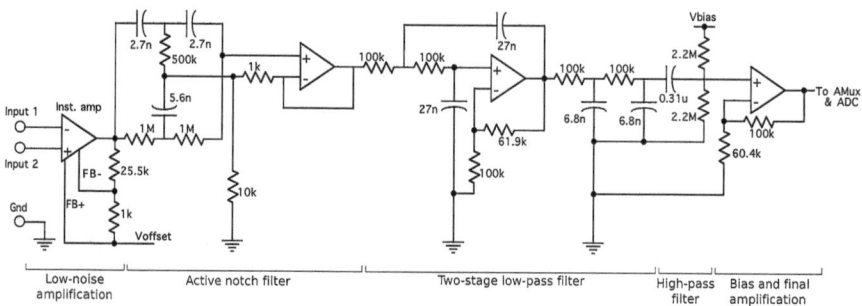

Fig. 3.39 An example of EEG AFE [3]

Exercise

Problem 3.1: For a thermistor with voltage output, the output at the lowest range of temperature is 0.2 V, while the output at the highest range of temperature is 0.3 V. The other end of the thermistor is connected to the mid-rail voltage of 1.5 V of a 3 V unipolar supply. Design (resistors from E-12 series) an inverting amplifier stage to maximize the swing of the 8-bit ADC input. Also determine the output voltage swing and the number of quantization steps it will produce.

    Problem 3.2: For a differential amplifier configuration from a Wheatstone Bridge, where the sensor resistance changes from 200 kΩ to 150 kΩ, design the circuit (E-12 series) for a 3 V supply. What is the output voltage swing? Also, determine the number of quantization steps for an 8-bit ADC.

    Problem 3.3: Design the schematic diagram of a 2-stage fourth-order Butterworth band-pass filter with 10 Hz center frequency and 1 Hz pass-band filter as shown below (produced with Analog Filter Wizard, Analog Devices) with a simulation SPICE software (e.g., KiCad).

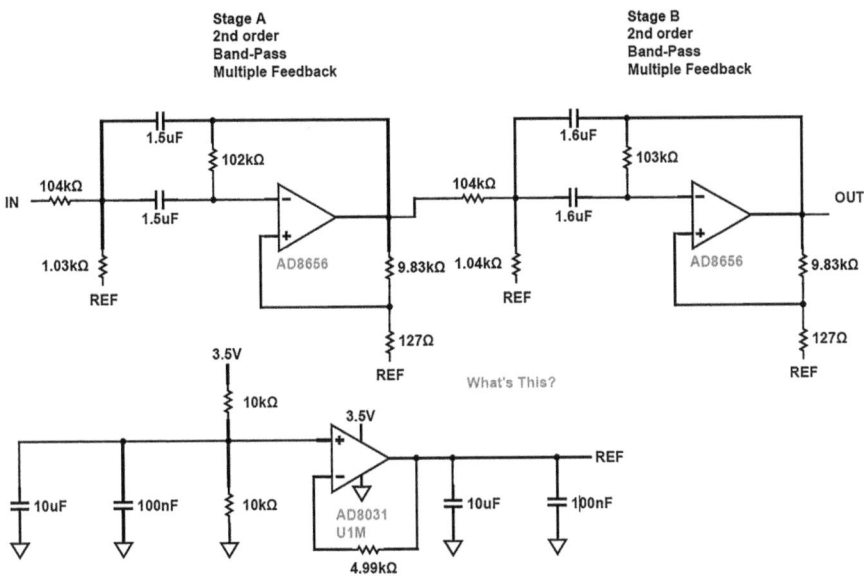

Simulate your design for 0.1, 1, and 10 Hz. Use idealopamp.cir for model file of the op-amps. Plot your results and show that this filter effectively passes 1 Hz compared to other two frequencies.

    Problem 3.4: For the 4-bit R-2R ladder in a 5 V system, find the analog output value when the digital input is 0110_b.

References

1. R. Mahajan, B.I. Morshed, Unsupervised eye blink artifact denoising of EEG data with modified multiscale sample entropy, kurtosis, and wavelet-ICA. IEEE J. Biomed. Health Inf. **19**(1), 158–165 (2015)
2. S. Khatun, R. Mahajan, B.I. Morshed, Comparative study of wavelet based unsupervised ocular artifact removal techniques for single channel EEG data. IEEE J. Transl. Eng. Health Med. **4**(1), 1–8 (2016)
3. S. Consul-Pacareu, R. Mahajan, M.J. AbuSaude, B.I. Morshed, NeuroMonitor: A low-power, wireless, wearable EEG device with DRL-less AFE. IET Circuits Devices Syst. J. **11**(5), 471–477 (2017)

Chapter 4
Software Design

> Design is not just what it looks like and feels like. Design is how it works.—Steve Jobs, co-founder of Apple Inc.

4.1 ES Software

Software is the program code that runs on a microprocessor or a microcontroller-based system to perform the desired tasks. Software needs to satisfy all objectives and specifications such as functionality, accuracy, stability, and I/O related operations. Embedded system composes of both hardware and software. Thus, an ES designer must also be adept on software design. However, unlike typical software programming on computers which is done in isolation of hardware consideration, embedded software design must be done with consideration of hardware. As Dr. Edward A. Lee of Berkeley summarized: *"Embedded software is software integrated with physical processes. The technical problem is managing time and concurrency in computational systems."*

A good ES designer will take advantage of the hardware resources available in the system and must meet hardware constraints such as real-time response with guaranteed timing latency. Figure 4.1 shows an example of an embedded system of a temperature sensing system where some portion of the information processing blocks are implemented in hardware (shown in gray rectangles) and the rest are implanted in software code (shown in white ellipses). Dr. Edward A. Lee identified that embedded software is software integrated with physical processes. The technical problem is managing time and concurrency in computational systems [1].

ES software design should be developed with not only functionality in mind, rather also consider quality of programming, efficiency and compactness of codes, and optimization both in software domain and hardware domain. There are two major categories of performance measures for software: qualitative and quantitative. Dynamic efficiency is a measure of how fast a program executes, and can be measured with number of CPU cycles or time in seconds. Static efficiency is the number of memory bytes required for a software code, which is measured in terms of global variables, stack space, fixed constraints, and program object codes.

© Springer Nature Switzerland AG 2021
B. I. Morshed, *Embedded Systems – A Hardware-Software Co-Design Approach*,
https://doi.org/10.1007/978-3-030-66808-2_4

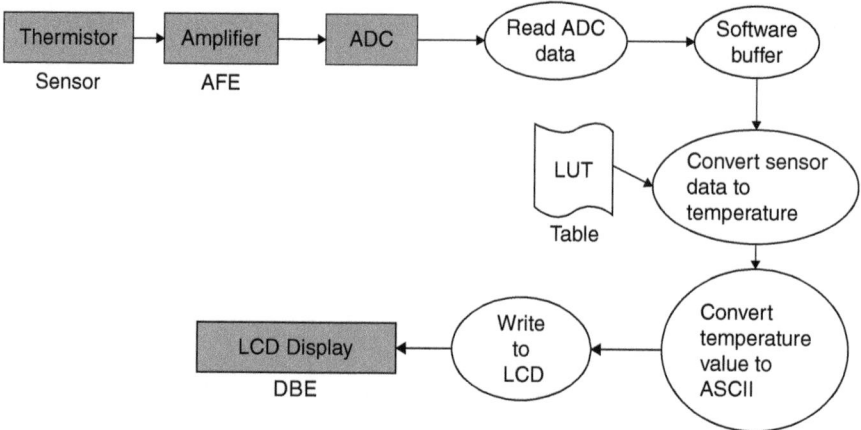

Fig. 4.1 An example of temperature sensing system with hardware (shown in gray) and software (shown in white) blocks

Today's ES are becoming complex, connected, and optimized. Consequently, software is also becoming complex with exponentially increasing complexity of the system. Modern ES systems can have more than 70% of the development cost for complex systems, such as automotive electronics, aviation, phones and other communication systems, are due to software development [2]. Thus, it is out-of-scope here to discuss all of the software choices, languages, and syntaxes for various software that can be used in ES design. Rather, here we briefly indicate some common programming language options and later provide some examples of C programming language with Arduino as the target platform. Readers must refer to relevant programming language books to learn these languages and syntaxes in details.

4.1.1   C Programming Language

C programming is one of the oldest programming languages that is still very popular for ES design due to its low-level hardware access capabilities, fast operation, and ability to run independently. This can be considered as Mid-level programming language. C program is converted to low-level assembly language with a compiler, then converted to machine code (i.e., binary file) with an assembler, along with help of linker if libraries are used. This machine code is downloaded on the MCU for execution of the program. As the machine code can run with or without operating system (OS), this is one of the best choices still for low-power ES design such as wearables or IoTs.

C coding is typically done with the MCU manufacturer provide or supported Integrated Development Environment (IDE). The file that contains the C code is called the source file, which is a text file. With the IDE integrated compiler and assembler, the source file is converted to assembly file, which is linked with other required precompiled files to an executable binary file. To download this binary file (hex) in an embedded processor, a programming chip is needed which might be integrated with the prototype board such as in Arduino or be an external device like In-circuit Debugger (ICD) of PIC microcontroller boards. The final hex file is downloaded in the microcontroller flash memory (which is non-volatile) and can be executed independently of operating system.

Although C program generates assembly codes, and ES programmer can also work on assembly code itself, however, this approach is becoming less popular with increasing complexity of codes. In cases of highly optimality need, such as execution clock cycle or memory usage, assembly code can still be useful to utilize. Most low-power embedded system, like Arduino, uses C programming. Thus, we will primarily focus on C here.

4.1.2   Java or Other High-Level Programming Language

Java is a high-level programming language. Many operating system (such as Android) support Java codes. It is a suitable programming language where the ES uses OS services such as file management, scheduling, and resource allocation. There are many other options for high-level programming languages such as Python, Perl, etc. As these programming languages are not efficient in implemented assembly code, they are not the best choice for efficient implementation. However, for complex codes such as smart devices with machine learning or deep learning techniques, these high-level programming language options are easier to implement. Nonetheless, ultimately these codes in high-level languages also are converted to assembly and machine code, which runs on the microcontroller hardware. Here, we will not focus on these high-level languages, but keep our discussion within C language, suitable for Arduino kit microcontrollers (such as Atmel MCU).

4.1.3   Pseudo-code

Pseudo-code is non-compile-able text representation of the code. It does not follow any exact syntax, and is meant to capture the imagination of the programmer to organize thoughts in common words. It is better suited when the software is more sequential and involves complex mathematical calculations. An example of pseudo-code:

```
Start
Read Name, Age
IF Age = -999 THEN
```

```
  Write "Error Empty File"
ELSE
  DOWHILE Name <> -999
    Write Name, Age
    Read Name, Age
  ENDDO
ENDIF
Stop
```

4.1.4 Flow-chart

Flow-chart is another intuitively understandable representation of the overview of the code. It also does not have any syntax, but typically follows some shape convention for different activities. It is good option when the software involves complex algorithm with many decision points causing control paths.

4.1.5 Assembly Programming Language

Assembly is a low-level programming language. It most closely represents machine code. Typically, each assembly code represents one or a few microprocessor clock cycles, which can be conclusively found from Instruction Set Architecture (ISA) manual. It is also the most efficient implementation of code, but becomes unmanageably difficult as the program becomes complex. Although historically MCUs were programmed with assembly language, newer MCU with complex code requirement and highly-efficient assembler that converts C code to smallest assembly code, use of assembly code is becoming rare. Sometimes C integrated assembly instructions are used. Furthermore, as assembly code is hardware specific, portability becomes a major challenge.

4.1.6 Quality of Programming

There are two major categories for quality considerations of programming: Quantitative performance measurements and Qualitative performance measurements. Quantitative performance measurement can be two types: Dynamic efficiency and Static efficiency. Dynamic efficiency is a measure of how fast a program can execute. It is typically measured in terms of seconds or processor clock cycles. Higher dynamic efficient codes will execute faster or require lesser number of clock cycles. Static efficiency is the number of memory bytes required. It is measured in terms of global variables, stack space, fixed constraints, and program object code. Smaller memory byte usage is preferred in ES as these MCUs are severely memory constrained.

4.1.7 Documentation and Commenting

It is very important to have codes documented well with comments. It makes it easier to understand if the code needs to be revisited later or to be shared with fellow programmer. The code below shows a commented code in C programming language:

```
// square root of n with Newton-Raphson approximation
r = n / 2; // Initial approximation
while ( abs( r - (n/r) ) > t ) { // Iteration loop
    r = 0.5 * ( r + (n/r) ); // Implementation of equation
}
System.out.println( "r = " + r ); // Printing of output
```

A golden rule to remember about commenting in software development is that: *"Write software for others as you wish they would write for you."*

4.1.8 Modular Software Design

One of the approaches for software development is modular design, where software development is divided into the software problem with distinct and independent modules. In this approach, the systems must be designed from components in modular fashion. A program module is a self-contained software task with clear entry and exit points. It must be easy to derive behavior from behavior of other subsystems. A module can be a collection of subroutines or functions that in its entirety performs a well-defined set of tasks. There are two requirement schemes: Specifications and Connections.

A critical issue in ES software is concurrency where multiple events are occurring in physical world which must be concurrently processed in ES. Modular software is a good mechanism to deal with this issue. Other major issues are synchronization and communication which can also be dealt with relatively easily with modular software design approach.

Some of the advantages of modular software design is easier to design, easier to change, and easier to verify. These are important for ES systems with complex nature. Communication of modules can take advantage of shared variables. This can be done with selecting the appropriate types of variables, such as private, public, local, global, static local, static global, and constant modifier.

4.1.9 Layered Software System

In another approach, software can be developed in a layered system. In layered approach, each layer of modules can call modules of the same or lower level, but not modules of higher level. This is a common approach used in operating system (OS). Here, the top layer is usually the main program layer or application layer. Other examples of layered architecture are Hardware abstraction layer (HAL), and Board-support package (BSP). In layered approach, a key concept used is abstraction, which is the process where data and programs are defined with a representation of similar to its meaning (semantics), while hiding away the implementation details.

4.2 ES Software Environments

Unlike traditional software in computers or servers which must have an OS to run the software, ES software can be deployed in environments with or without OS. Examples of OS-less software environment is Arduino (i.e., Atmel), ESP32, PIC, and PSoC. In fact, historic microcontrollers such as PAL, EEPROM, and FPGA, were always OS-less. With newer generation microprocessor-based micro-controller, there are many new options where an OS exists on the microcontroller that controls execution of the program. Examples of ES with OS are Raspberry Pi, Renesas Electronics Synergy, Beagle-Board, and NVidia Jetson.

4.2.1 MCU with OS

Operating system (OS) enables execution of multiple programs (using slots of processor times known as scheduling) and allocates system resources such as memory and peripherals. OS also handles access to peripherals and device drivers, as well as organizes storage such as file-system. It can also oversee access to other systems and manage communication protocols.

OS-based MCU are most suitable for complex program such as requiring audio/ video data processing, machine or deep learning, concurrent processor intensive tasks, and programs requiring file systems. An advantage of OS-based style is that various OS related services and libraries can be available, thus making it easy to access or manage the ports and memory. In this style, control and management is centralized by OS. In case of multitasking and scheduling, OS makes software processes to cooperate and share the processor time. The OS checks periodically if other process requires action or time, and if priorities or deadlines of the processes are maintained. However, OS-based programming requires careful programming skills to achieve these, and can lead to "buggy" process which can stall system.

OS for ES is quite different than traditional computer OS (i.e., standard OS). For instance, many OS for ES do not support disk, network, keyboard, monitor, or mouse by default, rather these functionalities can be added as-per-need by integrating tasks or libraries instead of integrated drivers. Effectively most of these devices do not need to be supported in all variant of OS for ES, except maybe the system timer which is generally required in most OS. Another major difference in OS for ES is that device drivers, middleware, and application software—all might directly access kernel, rather than traditional layered software architecture described earlier.

Many embedded systems are real-time systems, hence the OS used in these systems must be real-time operating system (RTOS). This RTOS is an operating system that supports the construction of real-time systems. The key requirements for RTOS are the timing behavior of the OS must be predictable, and OS should manage the timing and scheduling. An example of RTOS is RT-linux, which is commercially available from fsmlabs (www.fsmlabs.com).

Real-time tasks cannot use standard OS calls, as standard OS calls do not have any mechanism for timing constraint assertion. RTOS classifies various real-time tasks as Critical, Dependent, and Flexible. Critical tasks are those that must meet timing constraints. If the constraints are not met, the system fails. Dependent tasks can allow occasional failures to meet time constraints, but it is good to meet these constraints in most cases. These do not result in system failure; however, subsystems might be impacted. Flexible tasks timing constraints are for improved performance, but missed timing constraints do not cause any failure.

RTOS also must allow real-time scheduling. Scheduling is finding a mapping or a scheme to execute the given tasks such that constraints are met. Typically, these schedules have to respect a number of constraints, including resource constraints, dependency constraints, and deadlines. For example, if a given task graph $G = (V,E)$ is given, a schedule τ of G will be a mapping such that $V \rightarrow D_t$, where V is a set of tasks to start times with domain D_t so that all constraints are met.

4.2.2 OS-less

ES designer must ask the question if OS is needed for the project at hand. It is not essential! Sometimes OS is not needed for the required tasks, especially if the project is relatively simpler. Almost always, OS-less systems will consume less power for the same task compared to OS-based systems, as OS overhead power consumption is not required and OS-less MCUs can be put to deep-sleep state when no activity is required.

One common and easy approach to develop OS-less software program is to use "endless loop." The general structure of this are as follows:

```
loop:
  do the first task;
  do the second task;
```

```
...
do the n-th task;
wait before next round;
repeat loop;
```

The loop code typically precedes with an initialization and setup code that runs once every time the system is restarted or reset. Arduino code library (Sketch) is one of this type of endless loop by default. This style of coding is every efficient and predictable as there is no OS overhead or uncertainty. However, it may not be able to handle time constraints of asynchronous events and might waste clock cycles for various tasks when those are not required. This coding style suits nicely when the tasks are repetitive in periodic manner such as monitoring a vehicle speed.

One of the issues with endless loop is that it is not reactive and runs continuously even if the tasks are not required to be performed. Most recent ES systems such as IoTs and smart devices are generally reactive ES. For this type, another style might be more suitable and more optimal, which is "endless loop with interrupts." This style permits immediate action (real-time) for time constrained events. Interrupt task is performed with software codes known as interrupt service routine (ISR) which triggers when interrupts are registered. ISR is typically atomic, and thus should be kept small in terms of execution time. If more than one type of event is possible within the ISR execution time when the code was atomic, the subsequent event might be missed. This style of coding requires the programmer to be aware of and do event scheduling and guarantee timing deadlines.

4.3   ES Software System Considerations

For development of ES software needs to consider some specific considerations which are not typical for standard computer software development. Key aspects for these considerations are briefly described in this section.

4.3.1   ES System Classification

In terms of task requirements and predictability, ES systems can be classified in 3 types:

1. **Static:** The needs of the tasks of these systems are fixed and known at design time. Examples of this type of ES are traffic light system, elevator system, and oven control unit.
2. **Deterministic:** The needs of the tasks have variable requirements, but the requirements are either known or at least can be predicted fairly well in advance at design time. Examples of this type of ES are temperature control unit for air condition, drone control unit, and handheld camera controller.

3. **Dynamic:** The needs of the tasks have also variable requirements, but there may be no way to know or accurately predict these needs in advance at design time. Examples of this type of ES are intelligent robots, self-driving cars, and airplane control unit.

4.3.2 Real-Time Operation

Real-time with regards to computer system is defined as those systems that update information at the same rate as they receive data, enabling them to direct or control a process. Examples of real-time systems are air condition system controller, elevator controller, self-driving cars, multimedia player, etc. For real-time ES, the response must be relevant when it is generated. According to J. Stankovic [3], real-time systems are "*those systems in which the correctness of the system depends not only the logical results of the computation, but also on the time at which the results are produced.*"

Real-time systems will have some tasks that are real-time constrained. These tasks can be Hard real-time or Soft real-time. Hard real-time tasks are bounded by latency. They must be completed within this bound or else the whole system will fail. An example of hard real-time is braking of self-driving car. If the braking is not complete within certain time (or distance), the car might crash. Another example of hard-real time is flying controller of a drone. If there is a sudden gust of wind, the controller must compensate quickly or else it might crash land.

Soft-real time tasks needs to be executed as soon as possible. The sooner the task completes, the better the experience or performance. However, there is not strict timing or deadline that will result in system failure. An example of this type of system is the elevator control unit. Then a floor button is pressed, it is desirable for the elevator to reach the target floor. However, if it is little late, it does not cause any failure of system. Another example of this type is air condition control unit. When the temperature is too high or too low, air cooling or heating unit will turn on. It is desirable to complete this cooling or heating process soon, but delay of this process is not catastrophic to the system.

In addition, there can be tasks with no real-time constraints. They do not have any dependency of time. Although these types of tasks are rare in ES as most ES are real-time reactive systems. Real-time system design must consider real-time constraint. At design time, latencies (inherent delay) can be considered for timing constraints, but it needs to ensure the system delays at runtime also fits within the timing constraints, especially if it is a hard-real time constraint.

4.3.3 Real-time Reactive Operation

Real-time systems require correctness of result as a function of time when it is delivered. They do not have to be "real fast." In fact, in most cases, consistency is more important than raw speed. These systems should be tested for worst case performance (rather than typical performance). Worst case performance often limits design.

Reactive systems are those whose computation rate is in response to external events. If the events are periodic, they can be scheduled statistically. For example, the temperature monitoring of an air condition control unit can check the temperature once every one minute and take response if the sensed temperature is above or below the temperature range set by the user. For aperiodic events, however, the system must need to be able to statistically predict and dynamically schedule when possible. This will avoid overdesign. Another approach, which is superior, is to design for interrupt to trigger when the aperiodic events occur. For example, an RFID based lock system should only react when an RFID tag is presented, rather than checking for RFID tag periodically.

4.3.4 Time Services

Time plays a critical role in real-time systems. All microcontrollers have timer or counter hardware that can be used for relative timings (or delays). However, actual time is described by real numbers and not available in microcontrollers (unlike computers). If any project requires the exact time (i.e., year, month, day, hour, minute, and second), a Real-Time Clock (RTC) module must be included with the hardware. RTC can be initialized to current time, and can continue keeping time service. Constant power is needed to keep RTC counting, otherwise it will reset. Coin cell batteries are commonly used to only power the RTC that can operate for years. Two discrete standards are used in RTC: International Atomic Time (TAI) and Universal Time Coordinated (UTC). Both are identical on Jan. 1st, 1958. UTC is defined by astronomical standards and is more commonly used in ES. Timing behavior is essential for embedded systems. According Lee et al. (2005), the lack of timing in the core abstraction of computer science, from the prospective of ES, is an inherent flaw. Timing has far-reaching consequences for ES design processes.

4.3.5 Timing Specifications

According to Burns (1990), there are four types of timing specifications related to ES:

1. *Measure elapsed time:* This metric indicates how much time elapsed since last call.
2. *Means for delaying processes:* This metric indicates a procedure to delay a process relative to another process.
3. *Possibility to specify timeouts:* This specification provides a maximum time for a process to stay in a certain state. If the state does not go through a transition to next state within the timeout specification, timeout transitions to a different state (e.g., reset).
4. *Methods for specifying deadlines:* This specifies by which time a process must complete. For some processes, it might not exist, or may be specified in a separate control file.

4.4 Advanced Arduino Programming Techniques

As mentioned before, Arduino uses C programming language and is an OS-less software environment. It can either be used in an endless loop or an endless loop with interrupt. In this section, we will explore these with some advanced techniques in the context of Arduino Uno board which has Atmal ATMega328 microcontroller.

4.4.1 Bitmath

Bitwise manipulation provides great strength to programmer when faces with severely processing power constrained systems or extremely high-speed system. Some of the prime advantages of bit manipulation are:

1. When you need multiple pins to be set as input or output exactly at the same time (in the resolution of clock rate), using port bit manipulation allows it. For example, setting *PORTB &= B110011* output HIGH value to pins 8, 9, 12, and 13 at exactly the same clock cycle. Alternatively, if you use *DigitalWrite(8, HIGH); DigitalWrite(9, HIGH); DigitalWrite(12, HIGH); DigitalWrite (13, HIGH);* will require 4 instructions with slightly different time (in terms of clock cycle) of outputs. For high-speed time-sensitive cases (such as digital parallel communication), this time discrepancy might cause issues.
2. For high-speed communication (serial or parallel), bit manipulation is also powerful. These instructions only take a single clock cycle (typically), whereas *DigitalWrite* or *DigitalRead* instructions might require tens of instructions (depending on the library). In high-speed communication where you need to transmit data or receive data very quickly, bit manipulation is essential. Thus, the communication driver libraries utilize bit manipulation.
3. As bit manipulation requires lesser number of instructions, it is also helpful to reduce instruction code size, especially when the system has a very low program

memory (e.g., tiny microcontrollers) or power budget (e.g., battery or solar powered).

4.4.2 Arduino Uno Digital and Analog I/O Ports

To utilize Arduino Uno interrupt and timers, it is critical to understand and distinguish the onboard microcontroller (ATMega328) and board (Arduino Uno) I/O pins. While Arduino Uno board has 14 digital I/O pins (labeled 0–13) and 6 analog Input pins (labeled 0–5), the microcontroller refers these pins in 3 ports: Port B, Port C, and Port D. These are connected as per table below:

Arduino board I/O pins	Microcontroller I/O pins
Digital pins 0 to 7	Port D [0:7]
Digital pins 8 to 13	Port B [0:5]
Analog pins 0 to 5	Port C [0:5]

Note that Digital pins 0 and 1 are internally connect to RX and TX for serial communication used for USB connection to the computer for downloading the program code and serial port monitor/plotter. If this port is needed for connecting other hardware (such as Bluetooth module or voice recognition module), they must be connected exclusively (i.e., there must be only one communication connection at a certain time). For instance, if a sound recognition module needs to be connected, then one must remove the voice recognition module connection to program the board from the computer with USB cable. Then remove the USB cable, and connect the voice recognition module. As the board is typically powered by USB cable also, thus when USB cable is not connected, the Arduino Uno board must be powered otherwise (e.g., with a battery or a power adapter) using the DC power port.

Another note is the digital pin 13 is also internally connected to an onboard LED. Thus, whenever digital pin 13 is used as an output pin, the LED will correlate with the data being outputted. Also note that microcontroller Port B [6:7] and Port C [6:7] are not connected to any I/O port of the Arduino Uno board (rather used internally).

4.4.3 Bitmath Commands for Arduino Uno Digital I/O Pins

To setup a digital Arduino Uno I/O pin to input or output mode, the command used in Arduino Sketch library is *pinMode(pin,mode)*. This command sets that corresponding pin to INPUT or OUTPUT mode. For example, *pinMode(13,OUTPUT);* sets the digital pin 13 of Arduino Uno to output mode. In fact, the *pinMode* command writes 1 bit in the DDRx register, where x is based on the pin number as given in the table above. Alternatively, using Bitmath approach, the same operation can be done with writing the DDRx directly. The command will be *DDRx = value*.

Here x can be B, C, or D, depending on the target pin number. Binary value of 0 will set that pin as input, and 1 will set that pin as output. Note that DDRx command will write the all pins of that entire register (unless masking method is used as described below). Example: *DDRB = B100000;* command sets the pin 13 as output mode, and pin 8–12 as input mode. Note that "B" (upper case) on the right-hand side denotes the number is provided in binary format. Another notation for binary format is "0b" (smaller case).

To output a logic 0 (low) or 1 (high) using the output pin, Arduino Sketch code is *digitalWrite(pin, value)*. This command sets the pin to LOW (0) or HIGH (1). Example, *digitalWrite(13, HIGH);* will output high logic through pin 13 (and the onboard LED connected to pin 13 will turn on). In fact, this command writes one bit in the PORTx register to LOW or HIGH. Here, x is B, C, or D depending on the pin number. The Bitmath equivalent of this command is *PORTx = value*. This command sets the port bits to High (1) or Low (0) based on the value, which is reflected at the output pins. For example, *PORTB = B100000;* will set the pin 13 to HIGH (and pins 8–12 to low if set as output mode). Again, we can use masking technique described below if we do not want all bits of that port to change.

To read the digital value of an INPUT type pin, the Arduino Sketch code is *int value = digitalRead(pin);* that reads the data into the variable value. This command reads back the pin value (0 or 1). Example, *int data = digitalRead(7);* reads pin value from pin 7 (assuming it is set as Input). The Bitmath equivalent of data reading is a bit more complex. The command is *int value = PINx;* where x is Port B, C, or D depending on the pin. However, this command (PINx) reads the entire port and sets it in value. Thus, bit masking and shifting is essential after reading the port to get the correct value. Example, *int data = PIND;* reads the entire 8-bits of Port D to the variable data.

4.4.4   Bitmath Masking Techniques

Bitmath masking operations are typically done with AND, OR, or XOR gates. AND gate mask is used to Reset (i.e., logic 0) a specific bit. OR gate mask is used to Set (i.e., logic 1) a specific bit. XOR gate mask is used to toggle (i.e., make logic 1 if value was logic 0, and vice versa) a specific bit. To explain these operations, it is important to recall these gates Truth tables:

Input 1	Input 2	Output	Input 1	Input 2	Output	Input 1	Input 2	Output
0	0	0	0	0	0	0	0	0
0	1	0	0	1	1	0	1	1
1	0	0	1	0	1	1	0	1
1	1	1	1	1	1	1	1	0
AND gate truth table			OR gate truth table			XOR gate truth table		

AND gate example operation:
Initial portd is B01010011
Operation: **portd &= B00110101;**
Final portd is B00010001
Explanation:

Initial	0 1 0 1 0 0 1 1
AND (&)	0 0 1 1 0 1 0 1
Result	0 0 0 1 0 0 0 1

In general: **x & 0 → 0** and **x & 1 → x**, where x is the initial value.
OR gate example operation:
Initial portd is B01010011
Operation: **portd |= B00110101;**
Final portd is B01110111
Explanation:

Initial	0 1 0 1 0 0 1 1
OR (	)	0 0 1 1 0 1 0 1
Result	0 1 1 1 0 1 1 1

In general: **x | 0 → x** and **x | 1 → 1**, where x is the initial value.
XOR gate example operation:
Initial portd is B01010011
Operation: **portd ^= B00110101;**
Final portd is B01100110
Explanation:

Initial	0 1 0 1 0 0 1 1
XOR (^)	0 0 1 1 0 1 0 1
Result	0 1 1 0 0 1 1 0

In general: **x ^ 0 → x** and **~x ^ 1 → ~x**, where x is the initial value and ~x represents NOT(x).

4.4.5   Port Masking with Bitmath Examples

For further clarification, some examples of port masking with Bitmath operation are given below:

Example 4.1 What is the output for the operation below:

```
portb = portb & B000011;
```

Where the initial value of portb is B101010.

$$\begin{array}{lll}
\textbf{\textit{Solution:}} \text{ Initial portb:} & & 101010 \\
\text{Masking operator:} & \& & 000011 \\
& & ==== \\
\text{Final portb:} & & 000010
\end{array}$$

Explanation: This operation Reset the first 4 bits

Example 4.2   What is the output for the operation below:

```
portb = portb | B000011;
```

Where the initial value of portb is B101010.

$$\begin{array}{lll}
\textbf{\textit{Solution:}} \text{ Initial portb:} & & 101010 \\
\text{Masking operator:} & | & 000011 \\
& & ==== \\
\text{Final portb:} & & 101011
\end{array}$$

Explanation: This operation Set the last 2 bits

Note that the operation in example 4.1 can also be written as: *portb &= B000011;*. Similarly, the operation in example 4.2 can also be written as: *portb |= B000011;*.

4.4.6   Direct Port Access with Bitmath Examples

Here are some examples of direct port access with Bitmath beside the corresponding Arduino Sketch codes:

Arduino Sketch code	Bitmath equivalent code
pinMode(8,OUTPUT);	DDRB |= B000001;
digitalWrite(8,HIGH);	PORTB |= B000001;
pinMode(7,OUTPUT):	DDRD |= B100000000;
digitalWrite(7,LOW);	PORTD &= B011111111;
pinMode(9,INPUT);	DDRB &= B111101;
int val = digitalRead(9);	int val = (PINB & B000010) >> 1;

Note that the last code in this table in Bitmath column is performing a masking operation of the second bit (corresponding to digital pin 9), and subsequently shifting the value by 1 bit to the right to align that masked bit to LSB.

Table below gives more example for a partial code of Arduino:

Arduino Sketch code	Bitmath equivalent code
void setup () {	void setup () {
...	...
pinMode(5,OUTPUT);	ddrd |= B11100000;
pinMode(6,OUTPUT);	...
pinMode(7,OUTPUT);	}
...	void loop () {
}	...
void loop () {	portd |= B11100000;
...	...
digitalWrite(5,HIGH);	portd &= B00011111;
digitalWrite(6,HIGH);	...
digitalWrite(7,HIGH);	}
...	
digitalWrite(5,LOW);	
digitalWrite(6,LOW);	
digitalWrite(7,LOW);	
...	
}	

From the code snippet above, it is clear that the Bitmath can lead to a much smaller and compact code compared to Arduino Sketch code. Furthermore, each Arduino Sketch code statement might take several clock cycles to complete, whereas each Bitmath code statement takes only one clock cycle to complete. Thus, Bitmath operations, although more complicated to understand and use, leads to more static efficient code (requiring less memory) and more dynamic efficient code (requiring less clock cycles) for the exact same operation.

4.4.7 Advantages of Direct Port Access

Major advantages of using direct port access with Bitmath operations are:

1. **Execution time:** Turn pins ON or OFF very quickly (within the same clock cycle if they are in the same port or within a fraction of microseconds if different ports). On the other hand, *digitalRead()* and *digitalWrite()* get compiled into quite a few machine instructions. Each machine instruction requires one clock cycle, which can add up in time-sensitive applications. Direct port access can do the same job in a lot fewer clock cycles.
2. **Concurrent operation:** Bitmath operation can Set/Reset multiple output pins at exactly the same time (same clock cycle) when they are on the same port. Calling *digitalWrite(10,HIGH);* followed by *digitalWrite(11,HIGH);* will cause pin 10 to go HIGH several microseconds before pin 11. On the other hand, you can set both pins high at exactly the same moment in time using *PORTB |= B001100;* command.

3. *Compact code size:* Bitmath can make your code size smaller. It requires a lot fewer bytes of compiled code to simultaneously write via the port registers than to set each pin separately.

4.5   Dealing with Data Port Queue

Data port queue can be implemented using buffers. The purpose of buffer is to create a "spool" or a "queue" to allow the software routine to operate independently. There are two levels of buffering approach:

1. *Hardware level buffer:* This is typically a small (a few characters at most) buffer that is used directly by the subsystem to immediately obtain the data it demands or to immediately store the data it generates.
2. *Software level buffer:* This is typically a large buffer used when servicing the subsystem and is also used by the processing routine (main) to interact indirectly with the hardware

   Queueing mechanism is shown in Fig. 4.2.

4.5.1   Polling

The regular monitoring of a number of different service flag bits may be done through a technique called Polling. The polling technique consists of a loop which reads service flags and checks them one by one to see if they are active. Most peripherals provide *service flags,* which reflect I/O status conditions. Typically, these service flag bits get set automatically by the peripheral to indicate that a service routine needs to be launched. The service flags therefore need to be checked regularly to ensure timely service for some I/O task.

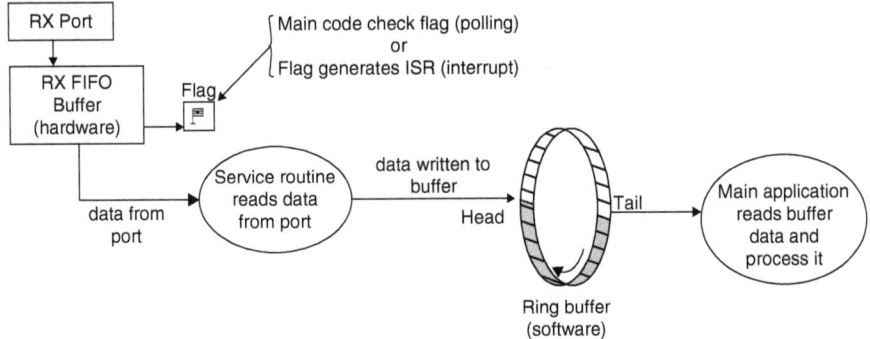

Fig. 4.2 Queueing mechanism for data access from MCU I/O port

The programmer needs to consider two things when service flags are used to launch service routines:

1. *Check the service flag bit regularly:* This involves periodic checking of the flag bit to see if new data is available at the port queue. Programmer needs to decide at the design time on how often the service flag needs to be checked. If the service flag become active for a short amount of time before new data arrive, the code might miss the first service flag and data loss might occur. If the service flag checking is too fast but data arrives after long intervals, then the clock cycles required to check the service flags are wasted overhead.
2. *Provide a specific service routine to serve the data:* When the service flag check finds the service flag bit active, it needs to quickly begin a service routine to serve the data (i.e., collect data from the port and move it to memory buffer or process data immediately). However, the delay might be unpredictable during design time as the exact time of arrival for data and time of checking for data is not known although bounded. Thus, the programmer needs to allow this uncertainty in the code. Also, the programmer needs to think in implementing priorities if multiple ports to be checked so that higher priority data are served appropriately within any timing deadline.

Thus, the polling technique has the following disadvantages:

- The latency interval between the time the service flag gets set and the time the program detects the active service flag depends on how busy the program is when the service request is made by the service flag.
- A large amount of processing time will be wasted polling service flags which are inactive.

4.5.2   Interrupt

An interrupt is an automated method of software execution in response to hardware event that is asynchronous with the current software execution. It allows program to respond to events when they occur and allows program to ignore events until that occurs. Interrupt technique is useful for external events, e.g., UART ready with/for next character, or signal change on pin. The action of the interrupt depends on context as programmed by Interrupt Service Routine (ISR). It can also be used to monitor number of edges (signal value change) arrived on pin. Interrupt can also be generated for internal events, e.g., power failure, arithmetic exception, or timer "tick" or overflow.

When an electronic signal causes an interrupt, the intermediate results have to be saved before the software responsible for handling the interrupt can run. This saving (and restoring process) takes some time, albeit small. The time delay incurred between the instant when the subsystem needs service to the instant the service routine actually provides the service is called *Interrupt Latency time*. Hardware buffer allows the system to cope with the "latency time" by providing a temporary storage. Key factors affecting latency are current CPU critical task, interrupt request processing, CPU registers that need to be saved in the stack, and initialization codes to begin servicing.

Some advantages of interrupts are as followings:

- Since the reaction to the service request is hardware based, there is no need to waste precious processing time polling inactive service flags.
- The latency interval to launch the interrupt handler will be very short when compared to polling.
- It will be easier to establish the priority structure which handles different service requests.

Some disadvantages of interrupt are as follows:

- The program necessary to control and interrupt system is much more complex than a polled system
- Hardware has to be added in order to connect the peripheral service flags to interrupt circuitry.
- Costs may be higher since the peripherals and the processor need additional hardware resources to support interrupt circuitry.

4.5.3 Software Buffer

Often individual data from ports needs to be placed to buffers when individual processing of data is not possible. For instance, to decode an audio signal, individual data samples cannot be processed. Similarly, to detect disease from ECG (electrocardiogram) data, it must be processed with at least one complete heartbeat representing ECG waveform. In some communication setup, especially wireless communication, data should be transmitted in chunks or packets, rather than individual data, as otherwise communication overhead becomes impractical.

For these scenarios, a software buffer as shown in Fig. 4.2 can be implemented. The buffer shown in this figure is "Ring buffer" or circular buffer. Another software buffer option is "Mutex buffer."

4.5.3.1 Ring Buffer

A schematic representation of ring buffer is shown in Fig. 4.3. The ring buffer has a write position (called Head) and a read position (called Tail). The data between Head and Tail are unprocessed data, and rest of the space is available. When new data is pulled from port, it will be placed at Head, and the Head position will be incremented. Similarly, when data needs to be processed, one or more data will be read from Tail position, and the new Tail position will be adjusted accordingly.

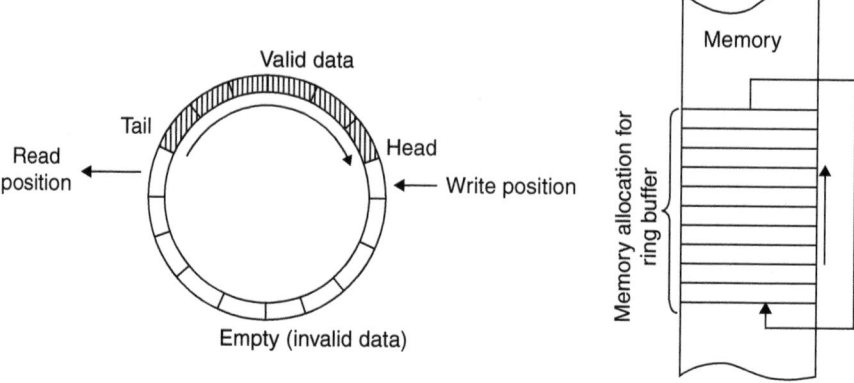

Fig. 4.3 Ring (or circular) buffer representation and implementation

There are a few constraints that needs to be met for Ring buffer, such as there must be available space for new data write (i.e., to increment Head position). As the Ring buffer will be implemented in memory with an allocation of memory, when Head (write process) reaches the top of the allocated memory, it must *Wraparound*, and begin from the bottom of the allocated memory. Similar wraparound must also happen to Tail when ready process reaches the top. Conditional checks must be performed to implement these.

4.5.3.2 Mutex Buffer

Although ring buffer is the most optimized software buffer, it is complex to implement. An easier solution is to employ Mutex buffer approach. Figure 4.4 shows a schematic representation of this approach. In this approach, a multiple of same size buffers are allocated. For instance, if 2-buffer space is used (A and B), the port will start writing data to one of them, say A. When A buffer is full, it can be used by the main code for processing. However, in the time A buffer is being processed, new data can still come in through input port. Thus, the new data needs to be written in buffer B. Again, when buffer B is full, the new data will be written in buffer A while buffer B data is being processed. This also needs to satisfy some constraints such as buffer size has to large enough to allow processing of data from the other buffer. More than 2 buffers can also be used in this approach.

4.5.4 Direct Memory Access (DMA)

Direct Memory Access (DMA) is useful to move large chunk of data to or from I/O port system from or to memory without involving MCU (except initialization).

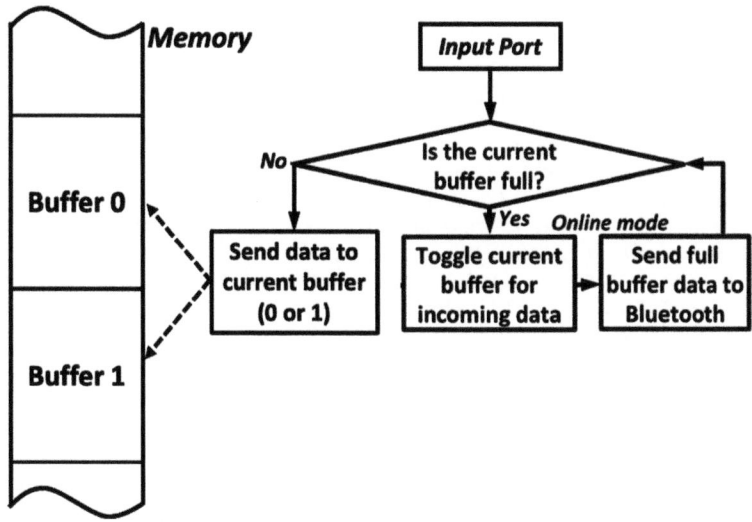

Fig. 4.4 Mutex buffer representation with 2-buffer space (A and B)

DMA controller is a hardware device that can control DMA data transfer without processor intervention. However, not all MCU contains DMA controller hardware. For example, the Arduino Uno microcontroller (i.e., ATmega328) does not have DMA hardware. But, Arduino Due MCU has DMA hardware. Most higher-end MCUs contain DMA controller hardware, such as MSP430 series MCUs, STMicroelectronics MCUs, and PSoC MCUs.

To setup DMA data transfer, the processor (MCU) needs to provide the following information to DMA controller:

- Beginning address in memory,
- Block length (i.e., number of Bytes to transfer),
- Direction (memory-to-port or port-to-memory),
- Port ID,
- End of block action (issue interrupt or do nothing).

After this setup process, DMA controller will begin data transfer, and MCU can perform other tasks or sleep, as per code. At the completion of DMA block transfer, DMA controller can raise an interrupt (wake up MCU) if it is set that way, or it can change corresponding status register to indicate completion of the data transfer task.

Example 4.3 Write a C code for transferring a block of ADC data to memory using DMA controller of Arduino Due using 4 mutex buffers.

Solution A code is provided below, where ADC data structure is assumed to consists of ADC_MR, ADC_CHER, ADC_IDR, ADC_IER, ADC_RPR, ADC_RCR, ADC_PNPR, ADC_RNCR, ADC_PTCR, ADC_CR, and ADC_ISR.

```
// For USB (ADC->DMA->Memory buffer; if full->USB)
#undef HID_ENABLED
// Input: Analog in A0
// Output: Raw stream of uint16_t in range 0-4095
// on Native USB Serial/ACM
volatile int bufn,obufn;
uint16_t buf[4][256]; // 4 buffers of 256 readings
void ADC_Handler(){   // move DMA pointers to next buffer
 int f=ADC->ADC_ISR; // Status register, Rx buffer (28)
 if (f&(1<<27)){ // Masking only 28th bit
  bufn=(bufn+1)&3; // Buffer sequence: 0->1->2->3->0 etc.
  ADC->ADC_RNPR=(uint32_t)buf[bufn];
 // Data from ADC to new Memory buffer
  ADC->ADC_RNCR=256; // Size
  }
}
void setup(){
 SerialUSB.begin(0);
 while(!SerialUSB); // Wait for USB ready
 pmc_enable_periph_clk(ID_ADC); // Enable peripheral clock
 // Initialize internal ADC module
 adc_init(ADC, SystemCoreClock, ADC_FREQ_MAX, ADC_STARTUP_FAST);
 ADC->ADC_MR |=0x80; // Mode for free running, no trigger
 ADC->ADC_CHER=0x80; // Enable Channel 7 of ADC
 NVIC_EnableIRQ(ADC_IRQn); // Enable interrupt
 // Disable all interrupt except channel 7
 ADC->ADC_IDR=~(1<<27);
 ADC->ADC_IER=1<<27; // Interrupt enable for channel 7
 // Initialize DMA buffer to begin with
 ADC->ADC_RPR=(uint32_t)buf[0];
 ADC->ADC_RCR=256; // Size of data
 // Initialize next DMA buffer to use
 ADC->ADC_RNPR=(uint32_t)buf[1];
 ADC->ADC_RNCR=256; // Size of data
 bufn=obufn=1; // Set current and old buffer as same
 // bufn: next buffer; obufn: overflow of next buffer
 ADC->ADC_PTCR=1;
 ADC->ADC_CR=2; // Control register to start DMA
 // write HIGH to bit-2 to start DMA operation
 }
void loop(){
 while(obufn==bufn); // wait for buffer to be full
 // send buffer data - 512 bytes = 256 uint16_t
 SerialUSB.write((uint8_t *)buf[obufn],512);
 obufn=(obufn+1)&3;   // select the next buffer
 }
```

Example 4.4 Write a C code for reading ADC data using DMA to store in a mutex buffer (2 buffers each of 1024 Bytes) and transfer entire buffer data from memory to serial port (Bluetooth) using DMA when buffer is full for a MSP430 MCU.

Solution A code is provided below.

```
// MSP430 uC; Read process from ADC12 using DMA
if (BF1_USE==1) // Which buffer to write data using DMA
  {
//src
  __data16_write_addr((unsigned short) &DMA0SA, &ADC12MEM0 );   //
dest
  __data16_write_addr((unsigned short) &DMA0DA, &Buffer2[0]);
   BF1_USE=0;
   BF1=1;
   } else {
//src
  __data16_write_addr((unsigned short) &DMA0SA, &ADC12MEM0 );
//dest
  __data16_write_addr((unsigned short) &DMA0DA, &Buffer1[0]);
  BF1_USE=1;
  BF2=1;
}
// BF1 = 1 -> Buf1 is full;
// BF2 = 1 -> Buf2 is full;
// x value increments until all done
 if (BF1==1) {
  UCB0TXBUF = *((unsigned char*)Buffer1+(x-1));
  x++;
  if (x==(1025)) {
   x=0;
   BF1=0; // Buf1 empty, transmission complete
   }
   } else if (BF2==1) {
  UCB0TXBUF = *((unsigned char*)Buffer2+(x-1));
  x++;
  if (x==(1025)) {
    x=0;
   BF2=0;  // Buf2 empty, transmission complete
   }
 }
}
```

4.6 Using Interrupts

To use interrupts, the peripheral must be instantiated with hardware capability in order to generate an interrupt request. Each peripheral is assigned its own interrupt request number. The initialization software for that peripheral must set the appropriate bits in the peripheral IRQ enable register. Hardware lines (IRQn) bring the different peripheral interrupt requests into the CPU where they are masked (using AND gates) by the iEnable control register. In order for a particular interrupt request to get through, the appropriate iEnable bit must be set by software when the system is initialized. Interrupt requests that get through the AND gates, apply 1's to the corresponding bit in the iPending control register. The iPending bits are then ORed together to generate a processor interrupt request which may be masked by the PIE bit located in the status control register. The PIE bit must be set by software when the system is initialized.

4.6.1   Interrupt Vectors

Interrupts use pre-programmed (ROM) Interrupt Vectors associated with the micro-controller. Interrupt Vector is a table in memory containing the first instruction of each interrupt handler. These are predefined interrupt routines to initiate recommended ISR functions. Programmer can also write custom ISR functions. If interrupts are not used, this memory can be used as part of the program. An example interrupt routine is RESET, which Sets up the stack pointer.

Interrupt exception routines service any interrupt. Before servicing the interrupt, all register values are saved in the stack. Then, the service is done through calling the appropriate Interrupt Handler. A function code called Interrupt service routine (ISR) starts to complete the service. After service is done, all registers are restored from the stack.

4.6.2   What Happens when Interrupt Event Occurs?

When an interrupt event occurs, the following operations takes place in sequence:

1. Processor does an automatic procedure call,
2. CALL automatically done to address for that interrupt,
3. Push current PC, Jump to interrupt address as each event has its own interrupt address,
4. The global interrupt enable bit (in SREG) is automatically cleared, i.e., nested interrupts are disabled, and
5. SREG bit can be set to enable nested interrupts if desired.
6. Interrupt procedure, aka "interrupt handler" or Interrupt Service Routine (ISR) starts,
7. ISR executes the function written inside the function,
8. ISR then returns via RETI,
9. The global interrupt enable bit is automatically set on RETI,
10. One program instruction is always executed after RETI.

4.6.3 ISR Coding Norms

Interrupt Service Routine (ISR) or interrupt handler should be written such that it is invisible to program, except through side-effects, e. g. via flags or variables. ISR will change the normal program execution timing (as ISR codes will need to be executed). Thus, interrupt-based program cannot rely on "dead-reckoning" using instruction timing only.

ISR needs to be written so they are invisible. For instance, ISR cannot stomp on program state, e. g. registers, or save and restore any registers used in the program. ISR code should be small and fast. So, do not put codes that takes long time to complete (such as serial port communication commands). It is also ideal to make ISR code "Atomic" to avoid firing of another ISR while one ISR is being executed.

4.6.4 List of Interrupts and ISRs of ATmega328

ATmega328 has a number of interrupts as shown below (taken from its datasheet [4]).

VectorNo.	Program address	Source	Interrupt definition
1	0×0000	RESET	External Pin, Power-on Reset, Brown-out Reset and Watchdog System Reset
2	0×0002	INT0	External Interrupt Request 0
3	0×0004	INT1	External Interrupt Request 1
4	0×0006	PCINT0	Pin Change Interrupt Request 0
5	0×0008	PCINT1	Pin Change Interrupt Request 1
6	0×000A	PCINT2	Pm Change Interrupt Request 2
7	0×000C	WDT	Watchdog Time-out Interrupt
8	0×000E	TIMER2 COMPA	Timer/Counter2 Compare Match A
9	0×0010	TIMER2 COMPB	Timer/Counter2 Compare Match B
10	0×0012	TIMER2 OVF	Timer/Counter2 Overflow
11	0×0014	TIMER1 CAPT	Timer/Counter1 Capture Event
12	0×0016	TIMER1 COMPA	Timer/Counter1 Compare Match A
13	0×0018	TIMER1 COMP8	Timer/Coutner1 Compare Match B
14	0×001A	TIMER1 OVF	Timer/Counter1 Overflow
15	0×001C	TIMER0 COMPA	Timer/Counter0 Compare Match A

(continued)

VectorNo.	Program address	Source	Interrupt definition
16	0×001E	TIMER0 COMPB	Timer/Counter0 Compare Match B
17	0×0020	TIMER0 OVF	Timer/Counter0 Overflow
18	0×0022	SPI, STC	SPI Serial Transfer Complete
19	0×0024	USART, RX	USART Rx Complete
20	0×0026	USART, UDRE	USART, Data Register Empty
21	0×0028	USART, TX	USART, Tx Complete
22	0×002A	ADC	ADC Conversion Complete
23	0×002C	EE READY	EEPROM Ready
24	0×002E	ANALOG COMP	Analog Comparator
25	0×0030	TWI	2-wire Serial Interface
26	0×0032	SPM READY	Store Program Memory Ready

A list of defined ISR of ATmega328 is below [4]:

#define	INT0_vect	_VECTOR(1)	/*	External Interrupt Request 0 */
#define	INT1_vect	_VECTOR(2)	/*	External Interrupt Request 1 */
#define	PCINT0_vect	_VECTOR(3)	/*	Pin Change Interrupt Request 0 */
#define	PCINT1_vect	_VECTOR(4)	/*	Pin Change Interrupt Request 0 */
#define	PCINT2_vect	_VECTOR(5)	/*	Pin Change Interrupt Request 1 */
#define	WDT_vect	_VECTOR(6)	/*	Watchdog Time-out Interrupt */
#define	TIMER2_COMPA_vect	_VECTOR(7)	/*	Timer/Counter2 Compare Match A */
#define	TIMER2_COMPB_vect	_VECTOR(8)	/*	Timer/Counter2 Compare Match A */
#define	TIMER2_OVF_vect	_VECTOR(9)	/*	Timer/Counter2 Overflow */
#define	TIMER1_CAPT_vect	_VECTOR (10)	/*	Timer/Counter1 Capture Event */
#define	TIMER1_COMPA_vect	_VECTOR (11)	/*	Timer/Counter1 Compare Match A */
#define	TIMER1_COMPB_vect	_VECTOR (12)	/*	Timer/Counter1 Compare Match B */
#define	TIMER1_OVF_vect	_VECTOR (13)	/*	Timer/Counter1 Overflow */

(continued)

#define	TIMER0_COMPA_vect	_VECTOR (14)	/ *	TimerCounter0 Compare Match A */
#define	TIMER0_COMPB_vect	_VECTOR (15)	/ *	TimerCounter0 Compare Match B */
#define	TIMER0_OVF_vect	_VECTOR (16)	/ *	Timer/Couner0 Overflow */
#define	SPI_STC_vect	_VECTOR (17)	/ *	SPI Serial Transfer Complete */
#define	USART_RX_vect	_VECTOR (18)	/ *	USART Rx Complete */
#define	USART_UDRE_vect	_VECTOR (19)	/ *	USART, Data Register Empty */
#define	USART_TX_vect	_VECTOR (20)	/ *	USART Tx Complete */
#define	ADC_vect	_VECTOR (21)	/ *	ADC Conversion Complete */
#define	EE_READY_vect	_VECTOR (22)	/ *	EEPROM Ready */
#define	ANALOG_COMP_vect	_VECTOR (23)	/ *	Analog Comparator */
#define	TWI_vect	_VECTOR (24)	/ *	Two-wire Serial Interface */
#define	SPM_READY_vect	_VECTOR (25)	/ *	Store Program Memory Road */

4.6.5   Interrupt Enabling and Disabling

By default, interrupt is disabled. Thus, the programmer needs to write code to enable it before using. For this, Global interrupt enable bit in SREG must be set. For Arduino, built in command *sei()* can be used that sets this bit. This allows all interrupts to be enabled with one bit. Similarly, *cli()* can be used to clear the bit, that disables all interrupts with one bit.

   Interrupt priority is determined by order in table. Lower addresses have higher priority. The function name format is: *ISR(vector)*, where vector function should define the interrupt routine. To return from ISR, *reti()* function can be used, however it is not required for built in ISR vectors as it is automatically generated for these ISRs.

4.6.6   Using External Interrupt of ATmega328

External interrupts monitor changes in signals on pins. To configure an interrupt, the corresponding control registers needs to be setup. Here, we will discuss two types of external interrupt briefly: INT and PCINT. Details of interrupt can be found in the datasheet [4].

The pins used in these external interrupts are listed below:

- INT0 and INT1 – range of event options

  - INT0 – PORT D [2] (i.e., Arduino digital pin 2)
  - INT1 – PORT D [3] (i.e., Arduino digital pin 3)

- PCINT[23:0] – any signal change (toggle)

  - PCINT[0:5] –> PORT B [0:5] (i.e., Arduino digital pins 8-13)
  - PCINT[8:13] –> PORT C [0:5] (i.e., Arduino analog pins 0-5)
  - PCINT[16:23] –> PORT D [0:7] (i.e., Arduino digital pins 0-7)

Note: PCINT for Port B, C, and D are referred to as PCINT0, PCINT1, and PCINT2, respectively. One of the constraints to use these interrupts is that Pulses on inputs must be slower than I/O clock rate.

4.6.7   INT Interrupts

To setup the control registers for INT interrupts, below is the information for setting up sense control bits for INT1 (INT0 sensor control is similar) [4].

Bit	7	6	5	4	3	2	1	0	
(0×69)	–	–	–	–	ISC11	ISC10	ISC01	ISC00	EICRA
Read/Write	R	R	R	R	R/W	R/W	R/W	R/W	
Initial value	0	0	0	0	0	0	0	0	

For the sense control of interrupt 1, see Table 12.1 to select ISC11 and ISC10 proper mode setting. For interrupt 0, similar mode selection needs to be done using ISC01 and ISC00. The external interrupt mask register is shown below. Write 1 to Enable, 0 to Disable corresponding interrupt in this control register.

Table 12.1 Interrupt 1 sense control

ISC11	ISC10	Description
0	0	The low level of INT1 generates an interrupt request.
0	1	Any logical change on INT1 generates an interrupt request.
1	0	The falling edge of INT1 generates an interrupt request.
1	1	The rising edge of INT1 generates an interrupt request.

Bit	7	6	5	4	3	2	1	0	
0×1D (0×3D)	–	–	–	–	–	–	INT1	INT0	EIMSK
Read/Write	R	R	R	R	R	R	R/W	R/W	
Initial value	0	0	0	0	0	0	0	0	

The external interrupt flag register is given below. Note that interrupt flag bit is set when a change triggers an interrupt request, where 1 represents interrupt is Pending, 0 represents no pending interrupt. Flag is cleared automatically when interrupt routine is executed, thus no explicit action is required when using ISR vectors. Also, flag can be manually cleared by writing a "1" to it.

Bit	7	6	5	4	3	2	1	0	
0×1C (0×3C)	–	–	–	–	–	–	INTF1	IMTF0	EIFR
Read/Write	R	R	R	R	R	R	R/W	R/W	
Initial Value	0	0	0	0	0	0	0	0	

For INT interrupts, programmer can choose to write their own custom ISR. To enable, use the instruction: *attachInterrupt(interrupt, function, mode);* where function can be used to write custom ISR function. Here, interrupt value is either 0 or 1 (for INT0 or INT1, respectively). The function represents the custom interrupt function to call. The mode value can be LOW, CHANGE, RISING, or FALLING. To disable the interrupt, use code: *detachInterrupt(interrupt);* where interrupt: Either 0 or 1 (for INT0 or INT1, respectively). Other related functions are *sei()* to enable global interrupt, and *cli()* to disable global interrupt.

4.6.8 PCINT Interrupts

The pin change interrupt (PCINT) control register is:

Bit	7	6	5	4	3	2	1	0	
(0×68)	–	–	–	–	–	PCIE2	PCIE1	PCIE0	PCICR
Read/Write	R	R	R	R	R	R/W	R/W	R/W	
Initial Value	0	0	0	0	0	0	0	0	

Here, PCIE2 enables interrupts for PCINT[23:16] (i.e., Port D), PCIE1 enables interrupts for PCINT[15:8] (i.e., Port C), and PCIE0 enables interrupts for PCINT [7:0] (i.e., Port B). The corresponding flag register is:

Bit	7	6	5	4	3	2	1	0	
0×1B (0×3B)	–	–	–	–	–	PCIF2	PCIF1	PCIF0	PCIFR
Read/Write	R	R	R	R	R	R/W	R/W	R/W	
Initial value	0	0	0	0	0	0	0	0	

Status flag of 1 when pending, and 0 when cleared. Note that the flag is cleared automatically when interrupt routine is executed by built in interrupt vector ISRs. The mask register for PCINT2 is given below:

Bit	7	6	5	4	3	2	1	0	
(0×6D)	PCINT23	PCINT22	PCINT21	PCINT20	PCINT19	PCINT18	PCINT17	PCINT16	PCMSK2
Read/ Write	R/W	R/W	R/W	R/W	R/W	R/W	R/W	R/W	
Initial value	0	0	0	0	0	0	0	0	

The bits 0 to 7 here corresponds to Arduino Pin 0 to 7 (Port D). Each bit controls whether interrupts are enabled for the corresponding pin. Note that change on any enabled pin causes an interrupt (not distinguishable). Mask registers 0 (PCMSK0) and 1 (PCMSK1) are similar.

Example 4.5 For an external interrupt (INT), complete the following partially filled up code.

```
#define pinint0
#define pinint1
     void setup()      (
        pinMode (pinint0,          );
        pinMode(pinint1,           );
        Serial.begin(9600);
        // External interrupts 0 and 1
        // Interrupt on rising edge
        EICRA =
        // Enable both interrupts
        EIMSK =
        // Turn on global interrupts
        sei ();
}
ISR(          _vect) {
}
ISR(          _vect) {
}
// Print out the information
void loop()
{
        Serial.print("X:  ");
        Serial.print(percent0);
        Serial.print("    Y:    ");
        Serial.println(percent1);
}
```

Solution
```
                #define pinint0 // Defined as Pin 2
   #define pinint1 // Defined as Pin 3
   int percent0 = 0;
   int percent1 = 0;
   void setup () {
    pinMode(pinint0, INPUT);
    pinMode(pinint1, INPUT);
    Serial.begin(9600);
    // External interrupts 0 and 1
    // Set interrupt on rising edge
    EICRA |= B00001111;
    // Enable both interrupts
    EIMSK |= B00000011;
    sei(); // Global Interrupt enable
   }
   ISR(INT0_vect) {
    percent0++;
   }
   ISR(INT1_vect) {
    percent1++;
   }
   // Print out the information
   void loop () {
    // Put MCU to sleep to save power
    // See subsequent slides for details
    Serial.print("X: ");
    Serial.print(percent0);
    Serial.print("  Y: ");
    Serial.println(percent1);
   }
```

Example 4.6 Setup your hardware as follows with an Arduino Uno board:

- From VDD to two push switches in series with two resistors (1 kΩ each).
- Connect Pin 8 to one switch at resistor
- Connect Pin 3 to the other switch at resistor

Write a code that uses PCINT such that when one of the switches connected to Pin 8 is pressed, the *value* increments, but when the switch connected to Pin 5 is pressed, the *value* decrements. Use serial port monitor in Arduino sketch to monitor this data.

Solution A code is provided below to perform the above task.

```
   #include <avr/interrupt.h>
   volatile int value = 0;
   void setup () {
    // Global interrupt disable - Atomic
    cli();
```

```
// Enable PCINT0 and PCINT2
PCICR |= B00000101;
// Mask for Pin 8 (Port B)
PCMSK0 |= B00000001;
// Mask for Pin 3 (Port D)
PCMSK2 |= B00001000;
// Global interrupt enable
sei();
// Serial port initialize
Serial.begin(9600);
}
// ISR for Pin 8 interrupt, inc value
ISR(PCINT0_vect) {
 value++;
}
// ISR for pin 3 interrupt, dec value
ISR(PCINT2_vect) {
 value--;
}
// Main loop prints value
// Note: no port checking statement
void loop () {
 Serial.println(value);
}
```

Example 4.7 How can you use interrupt for implementing a ring buffer?

Solution Interrupt can be used to monitor input data ports and the ISR will collect data from the port to the ring buffer. The Head pointer will be incremented for each data read. In the main code, data from ring buffer will be read and corresponding Tail pointed will be moved accordingly. Boolean variables can be used to check the constraints such as overflow condition. A partial code is provided below. Here one block of data can be processed by the main code, thus it must wait until enough data is collected.

```
// Main code --- Read process
...
if (Overflow == 0) {
   Valid_Data = Head - Tail;  // Head & Tail as defined
 } else {
   Valid_Data = Head + buffer_size - Tail; // Wraparound
}
if (Valid_Data >= block_size)
   {
 (read data block)
...
```

```
// Check if sufficient data of 1 block has accumulated
if ((Tail + block_size) < buffer_size)
     Tail = Tail + block_size;
    else{
    Overflow = 0;
    Tail = Tail + block_size - buffer_size;
       }
...
// ISR code ---- Write process
...
if(Head == buffer_size){ // Wraparound
    Head = 0;
    Overflow = 1;
  }
// detect error condition
// Tail > Head for Overflow = 1
  if((Overflow == 1) && (Tail < Head))
  hold_Flag = 1; // halts new data to put in Ring buffer
  else
  hold_Flag = 0; //allows new data to put in ring buffer
 ...
  if((Head < buffer_size) && (hold_Flag == 0)){
...
// reads data from analog pin 0
    buffer[Head++] = analogRead(0);
 ...
```

4.7   Sleep Modes

One of the major advantages of OS-less MCU is the use of deep-sleep modes that drastically reduce power consumptions. However, to wake up the MCU from deep sleep, interrupt needs to be used (leading to reactive ES). Now that we have studied interrupt, it is a good time to look into sleep modes. Here we will focus on ATmega328 microcontroller sleep modes as this is the MCU of Arduino Uno board.

4.7.1   ATmega328 Sleep Modes

The table below shows various sleep modes available in this MCU [4].

Sleep mode	Active clock domains					Oscillators		Wake-up sources								Software BOD Disable
	Clk_CPU	Clk_FLASH	Clk_IO	Clk_ADC	Clk_ASY	Main Clock Source Enabled	Timer Oscillator Enabled	INT1, INT0 and Pin Change	TWI Address Match	Timer2	SPM/ EEPROM Ready	ADC	WDT	Other/ O	
Idle			x	x	x	x	x[b]	x	x	x	x	x	x	x	
ADC Noise Reduction				x	x	x	x[b]	x[c]	x	x[b]	x	x	x		
Power-down								x[c]	x				x		x
Power-save					x		x[b]	x[c]	x	x			x		x
Standby[a]						x		x[c]	x				x		x
Extended Standby					x[b]	x	x[b]	x[c]	x	x			x		x

[a]Only recommended with external crystal or resonator selected as clock source.
[b]If Timer/Counter2 is running in asynchronous mode.
[c]For INT1 and INT0, only level interrupt.

Note that some hardware units are turned off in some sleep modes. This is critical to understand to use the sleep modes. The objective is to use the lowest power sleep mode provided the needed hardware during the sleep mode is active. ATmega328 microcontroller typical current consumptions in various sleep conditions (high to low) are given below with a list of key hardware units that are available in that sleep mode:

SLEEP_MODE_IDLE	15 mA	All I/O, clk, timers, mem
SLEEP_MODE_ADC	6.5 mA	ADC, EEPROM, clk, Timer2, mem
SLEEP_MODE_PWR_SAVE	1.6 mA	Main clk, clk(asy), Timer2
SLEEP_MODE_EXT_STANDBY	1.6 mA	Clk(asy), Timer osc, Timer2 (no main clk)
SLEEP_MODE_STANDBY	0.8 mA	Main clk
SLEEP_MODE_PWR_DOWN	0.4 mA	Everything off (expt. Interrupt, WDT)

4.7.2 How to Enable Sleep Mode?

To enable sleep mode, the avr/sleep.h library needs to be included. In the setup code, write instruction for *set_sleep_mode(mode);* that initializes the sleep mode. In the loop function, write *sleep_mode();* to active sleeping. This statement is actually a combination of 3 statements in sequence: *sleep_enable(); cpu_sleep(); sleep_disable ();*

```
#include <avr/interrupt.h>
#include <avr/sleep.h>
void setup() {
  ...
  set_sleep_mode(SLEEP_MODE_IDLE); // select mode
  ...
}
void loop () {
  ...
  sleep_mode();
  ...
}
```

When an interrupt occurs to wake up the MCU, it resumes from the sleep_mode (or sleep_disable) statement.

Example 4.8 Setup your Arduino Uno with hardware as shown below. Notes for hardware setup:

- From VDD (3.3V), connect a push switch and a 1kΩ resistor in series to ground (0V).
- Pin 10 connects to the midpoint of the push switch and 1kΩ resistor.

- From pin 13, connect an LED (red) with a 220Ω (or 330Ω) resistor in series to ground.
- The 4-pin push switch has two sets of pins internally connected. To connect the switch properly, connect the two pins on the same side of the switch.

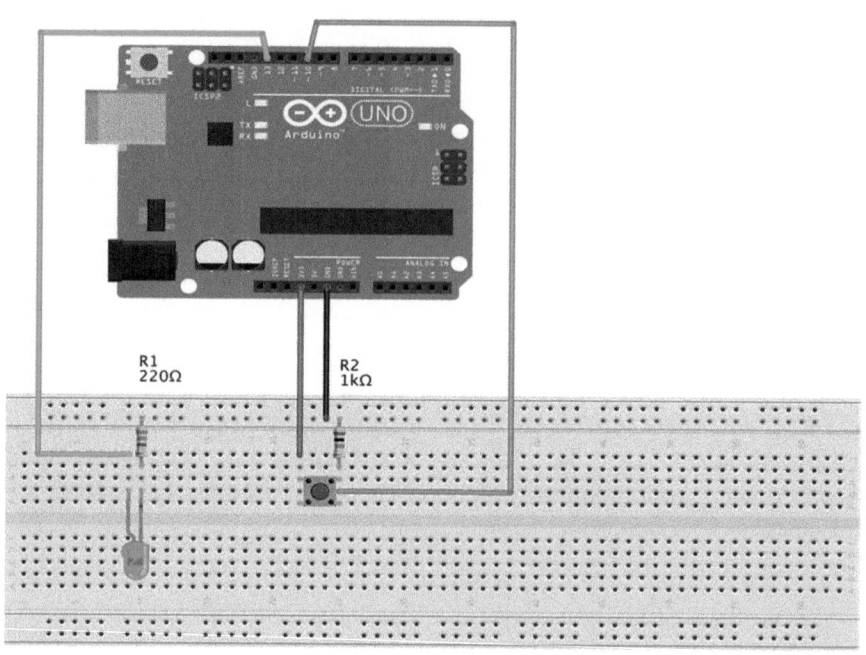

Now complete the partially filled up code below, so that the LED toggles when the push switch is pressed. Use the highest possible sleep mode (i.e., lowest current during sleep).

```
#include <avr/interrupt.h>
#include <avr/sleep.h>
void setup() {
  cli(); // Clear global interrupt
  // Set Pin 13 as output and 10 as input
  DDRB |=          ;
  DDRB &=          ;
  // Control regs for PCINT
  PCICR |=         ; // Enable PCINT0
  PCMSK0 |=        ; // Select PCINT0 mask
  // Serial.begin(9600); // Only for debug
  sei(); // Set global Interrupt
  // Use an appropriate sleep mode
  set_sleep_mode(      );
}
```

```
// ISR for pin change interrupt capture
// Note: triggers both on rising & falling
ISR(      _vect) {
 // Display in serial monitor for debug
 // Serial.println("Switch pressed");
 // Toggle the LED
 PORTB      ;
}
// Main loop
void loop() {
 // Display in serial monitor for debug
 // Serial.println("Main loop");
 // Do nothing!
 // Put MCU to sleep
}
```

Solution The completed code is given below:

```
#include <avr/interrupt.h>
#include <avr/sleep.h>
void setup() {
  cli(); // Clear global interrupt
  // Set Pin 13 as output and 10 as input
  DDRB |= B10000000 ;
  DDRB &= B111011   ;
  // Control regs for PCINT
  PCICR |= B00000001 ; // Enable PCINT0
  PCMSK0 |= B00000100 ; // Select PCINT0 mask
  // Serial.begin(9600); // Only for debug
  sei(); // Set global Interrupt
  // Use an appropriate sleep mode
  set_sleep_mode(SLEEP_MODE_PWR_DOWN);
}
// ISR for pin change interrupt capture
// Note: triggers both on rising & falling
ISR(PCINT0_vect) {
 // Display in serial monitor for debug
 // Serial.println("Switch pressed");
 // Toggle the LED
 PORTB ^= B100000 ;
}
// Main loop
void loop() {
 // Display in serial monitor for debug
 // Serial.println("Main loop");
 // Do nothing!
 // Put MCU to sleep
 sleep_mode();
}
```

4.8   Using MCU Internal Timer/Counter Hardware Unit

Precise time (delay) count requires use of hardware timer/counter internal to the MCU. It is also useful for timed ADC data capture. Never use *for loop* to generate delay in ES, as it is the most inefficient method. Arduino Sketch built in *delay* function uses these internal timer/counter hardware, which is great for precision timing. But MCU cannot be put to sleep with the *delay* function. Thus it is essential for understanding and using the timer/counter hardware directly to take advantage of sleep capabilities of MCU.

4.8.1   Internal Timer/Counter of ATmega328

ATmega328 has three timer/counter units as follows:

- Timer/counter0: 8-bit
- Timer/counter1: 16-bit
- Timer/counter2: 8-bit

Note: Only Timer/counter2 is ON during sleep modes (except IDLE mode). Below is the list of registers related to timer/counter of ATmega328.

TCNTx—Timer/counter count register
OCRxA—Output Compare Register A
OCRxB—Output Compare Register B
TCCRxA—Timer/counter control register A
TCCRxB—Timer/counter control register B
TIMSKx—Timer/counter interrupt mask register
TIFRx—Timer/counter interrupt flag register

Here, x can be 0, 1, or 2 for corresponding timer/counter. The timers/counters can generate two types of interrupts: TOV for overflow, and Computer A&B types for comparing to a set value. In this text, we will learn about TOV as this is simpler to code.

The timer/counter control registers for timer/counter0 are given below. Control registers of Timer/Counter1 and Timer/Counter2 are similar.

Bit	7	6	5	4	3	2	1	0	
0×24 (0×44)	COM0A1	COM0A0	COM0B1	COM0B0	–	–	WGM01	WGM00	TCCR0A
Read/ Write	R/W	R/W	R/W	R/W	R	R	R/W	R/W	
Initial value	0	0	0	0	0	0	0	0	

Bit	7	6	5	4	3	2	1	0	
0×25 (0×45)	FOC0A	FOC0B	–	–	WGM02	CS02	CS01	CS00	TCCR0B
Read/Write	w	W	R	R	R/W	R/W	R/W	R/W	
Initial value	0	0	0	0	0	0	0	0	

These two control registers setup the counter mode using WGM (Waveform Generation Mode) bits and pre-scaler using Clock Source select bits. We will use normal mode for WGM, as given in the table below.

Mode	WGM02	WGM01	WGM00	Timer/counter mode of operation	TOP	Update of OCRx at	TOV Flag Set on[a,b]
0	0	0	0	Normal	0×FF	Immediate	MAX
1	0	0	1	PWM, Phase Correct	0×FF	TOP	BOTTOM
2	0	1	0	CTC	OCR A	Immediate	MAX
3	0	1	1	Fast PWM	0×FF	BOTTOM	MAX
4	1	0	0	Reserved	–	–	–
5	1	0	1	PWM, Phase Correct	OCRA	TOP	BOTTOM
6	1	1	0	Reserved	–	–	–
7	1	1	1	Fast PWM	OCRA	BOTTOM	TOP

[a]MAX = 0×FF
[b]BOTTOM = 0×00

As these timers are small (8 or 16 bits), they cannot count long periods if driven directly my internal clock. To provide a better solution, these timer/counter can use pre-scalar hardware that can divide the internal clock by 1 to 1,024 factors by Clock Source (CS) select bits. For CS selection table for Timer/Counter0 is given below. Timer/Counter1 table is similar.

CS02	CS01	CS00	Description
0	0	0	No clock source (Timer/Counter stopped)
0	0	1	$Clk_{I/O}$/(No prescaling)
0	1	0	$clk_{I/O}$/8 (From prescaler)
0	1	1	$clk_{I/O}$/64 (From prescaler)
1	0	0	$clk_{I/O}$/256 (From prescaler)
1	0	1	$clk_{I/O}$/1024 (From prescaler)
1	1	0	External clock source on T0 pin. Clock on falling edge
1	1	1	External clock source on T0 pin. Clock on rising edge

Here the external clock can be applied with T0 pin which is connected to Port D [4], and T1 pin which is connected to Port D[5]. The CS selection table for Timer/Counter2 is different as given below. Note that this timer/counter does not have external clock source selection option.

CS22	CS21	CS20	Description
0	0	0	No clock source (Timer/Counter stopped)
0	0	1	clk_{T2S}/(No prescaling)
0	1	0	clk_{T2S}/8 (From prescaler)
0	1	1	clk_{T2S}/32 (From prescaler)
1	0	0	clk_{T2S}/64 (From prescaler)
1	0	1	clk_{T2S}/128 (From prescaler)
1	1	0	clk_{T2S}/256 (From prescaler)
1	1	1	clk_{T2S}/1024 (From prescaler)

Output of the timer/counter can be configured to an output pin. Thus, we can use software to generate clock signal of (almost) any frequency. The interrupt mask register for Timer/Counter0 is:

Bit	7	6	5	4	3	2	1	0	
(0×6E)	–	–	–	–	–	OCIE0B	OCIE0A	TOIE0	TIMSK0
Read/Write	R	R	R	R	R	R/W	R/W	R/W	
Initial Value	0	0	0	0	0	0	0	0	

Here, TOIE0 allows Timer Overflow Interrupt Enable, and OCIE0A/B is to setup Compare A/B interrupt enable. The corresponding flag register is:

Bit	7	6	5	4	3	2	1	0	
0×15 (0×35)	–	–	–	–	–	OCF0B	OCF0A	TOV0	TIFR0
Read/Write	R	R	R	R	R	R/W	R/W	R/W	
Initial Value	0	0	0	0	0	0	0	0	

Where TOV0 is for Timer overflow flag, and OCF0A/B is for Compare A/B interrupt flag. Timer/Counter 1 and 2 have similar registers.

4.8.2 OVF Timer Interrupt

To use OVF (Overflow) timer interrupt, we need to set WGM to Normal mode (i.e., 0). In this configuration, timer value increments, and when it reaches the highest value (i.e., 0xFF for 8-bit and 0xFFFF for 16-bit), it wraps around at TOP. At that instance, OVF interrupt is issued. Note that the timer/counter starts again at 0, so must reset to counter value if count must be done for a smaller value that the full range. For example, if using Timer/Counter0, then TOV0 interrupt flag will be set when TCNT0 resets to 0x00 from 0xFF. It can be used to generate periodic signals (like a clock). Note that Timer/Counter2 can be used with sleep mode to develop ultra-low-power ES like battery-powered IoTs. Timers can also be used to generate an interrupt every N time units. For this purpose, set TCNT0/2 to an initial value of (255-N), or for the case of TCNT1, an initial value of (65,535-N).

Example 4.9 What is the value of Timer1 needed for 1 second count in ATmega328 with 16 MHz clock?

Solution Internal clock $= 16$ MHz. As the needed time is large, we will use the maximum value of pre-scaler, which is 1024. Thus, clock cycles needed $= 16 * 10^6/ 1024 = 15{,}625$. Timer1 is 16-bit, so maximum value (i.e., 0xFFFF) $= 65{,}535$. So, Timer1 count value needed $= 65{,}535 - 15{,}625 = 49{,}910$. Thus, the value of Timer1 needs to be set to 49,910 or 0xC2F6.

Example 4.10 Write a code for 1 second timer for ATmega328 (Arduino Uno). The hardware setup: connect pin 13 to an LED in series of 220 Ω or 330 Ω resistor to ground.

Solution Code for 1 second timer using Timer/Counter1 hardware.

```
void setup () {
 cli(); // Disable global interrupt - atomic
 DDRB |= B100000; // Pin 13 output
 // Set timer 1 to normal mode
 TCCR1A = B00000000;
 // Set pre-scaler to 1024
 TCCR1B = B00000101;
 // Turn ON OVF
 TIMSK1 = B00000001;
 // Initial Timer1 value for 1 sec count
 TCNT1 = 0xC2F6;
 sei(); // Enable global interrupt
}

// Timer1 ISR
ISR(TIMER1_OVF_vect) {
 // Toggle output pin each 1 sec
 PORTB ^= B100000;
 // Reset counter value for next 1 sec
 TCNT1 = 0xC2F6;
}
// Main loop
void loop () {
 // Do nothing
 // If including sleep mode, ensure
 // timer1 is ON while sleep
}
```

Example 4.11 Write a code for 1 second timer for ATmega328 (Arduino Uno) such that a low-power sleep mode can be used (other than IDLE mode). The hardware setup: connect pin 13 to an LED in series of 220 or 330 Ω resistor to ground.

Solution Code for 1 second timer using Timer/Counter2 hardware.

```
#include <avr/sleep.h>
char rep = 0; // Timer repeat count
void setup () {
```

```
// Set pin 13 as output
DDRB |= B100000;
// Using Timer2, normal mode
TCCR2A = B00000000;
// Pre-scaler 1024 (max)
TCCR2B = B00000111;
// Pre-scaled clock rate = 16M/1024
// = ~16k
// Timer max count=16k/0.25k=~64
// Turn on OVF interrupt
TIMSK2 = B00000001;
// Turn on global interrupt
sei();
}
// ISR for TOV that triggers it
ISR(TIMER2_OVF_vect) {
 rep++; // Increment repeat count
 // For 1 sec, 64 repeats needed
 if (rep == 64) {
  rep = 0; // Reset repeat count
   PORTB ^= B100000; // toggle bit 13
 }
}
// Main loop
void loop() {
 // void loop Nothing to do
 // set sleep mode and sleep cpu
 set_sleep_mode(SLEEP_MODE_PWR_SAVE);
 sleep_mode();
}
```

Example 4.12 Complete the partial code for 1/64 second timer interrupt output generator for ATmega328 (Arduino Uno).

```
char timer = 0;
void setup() {
 DDRB =       ; // Pin 13 as output
 // Using timer, Set to Normal mode, Pin OC0A disconnected
 TCCR2A =     ;
 // Prescale clock by 1024, Interrupt every 256K/16M sec = 1/64 sec
 TCCR2B =     ;
 // Turn on timer overflow interrupt flag
 TIMSK2 =   ;
 sei(); // Turn on global interrupts
}
ISR(    _vect) {
 timer++;
 PORTB =       ; // Toggle bit 13
}
void loop() {
 // Do nothing
}
```

Solution   The completed code is provided below.

```
char timer = 0;
void setup() {
 DDRB |= B100000  ; // Pin 13 as output
 // Using timer, Set to Normal mode, Pin OC0A disconnected
 TCCR2A = B00000000 ;
 // Prescale clock by 1024, Interrupt every 256K/16M sec = 1/64 sec
 TCCR2B = B00000111 ;
 // Turn on timer overflow interrupt flag
 TIMSK2 = B00000001 ;
 sei(); // Turn on global interrupts
}
ISR(TIMER2_OVF_vect) {
 timer++;
 PORTB ^= B100000 ; // Toggle bit 13
}
void loop() {
 // Do nothing
}
```

Example 4.13   Write a code to generate a 100 Hz waveform through Pin 13 using timer2 hardware of ATmega328 (Arduino Uno).

Solution   Using max pre-scaler, modified clock $= 16M/1024 = 15{,}625$ Hz. For 100 Hz, we need 10 ms clock period, i.e., 5 ms ON time, 5 ms OFF time. For 5 ms toggle timer, number of modified clock cycles$=15{,}625 * 5 * 10^{-3} = {\sim}78$. Count value needed for Timer $2 = 255 - 78 = 177 = $ 0xB1.

The complete code is provided below.

```
void setup () {
 // Set pin 3 as output (arbitrary)
 DDRD |= B00001000;
 // Using Timer2, normal mode
 TCCR2A = B00000000;
 // Pre-scaler for 1024
 TCCR2B = B00000111;
 // Turn on OVF interrupt
 TIMSK2 = B00000001;
 // Set initial count value
 TCNT2 = 0xB1;
 // Turn on global interrupt
 sei();
}
// ISR for TOV that triggers it
ISR(TIMER2_OVF_vect) {
 // Toggle output: On->Off->On etc.
 PORTD ^= B00001000; // toggle bit 3
 // Re-initialize timer count value
 TCINT2 = 0xB1;
}
// Main loop
```

```
void loop () {
  // Nothing to do
  // CPU can be put to sleep with
  // a proper mode selection
}
```

Example 4.14 Write a code for a traffic light control hardware (Red, Yellow, Green light outputs, and a pedestrian push switch input) using timer hardware of ATmega328 (Arduino Uno).

Solution   A complete code is provided below.

```
#include <avr/interrupt.h>
#include <avr/sleep.h>
char tick = 0; // Unit time
void setup () {
  cli ();
  // Set pin 11, 12, 13 as output
  DDRB |= B111000;
  // Set pin 10 as input ped switch
  DDRB &= B111011;
  // Timer 1 code, normal mode
  TCCR1A = B00000000;
  // Pre-scaler 1024 (max)
  TCCR1B = B00000101;
  // Turn on OVF interrupt
  TIMSK1 = B00000001;
  // Initialize for 1 sec time tick
  TCNT1 = 0xC2F6;
  // setup input switch interrupt: PCINT0
  PCICR |= B00000001;
  // Pin 10 as input for PCINT0 interrupt
  PCMSK0 &= B111011;
  // For debugging only
  // Serial.begin (9600);
  // Turn on global interrupt
  sei ();
}
// ISR for TOV that triggers it
ISR (TIMER1_OVF_vect) {
  tick++; // Increment tick -> elapsed sec
  switch(tick) {
    case 5: PORTB |= B001000; break; // Yellow
    case 6: PORTB |= B010000; break; // Green
    case 10: PORTB |= B100000;
            tick = 0; break; // Red, reset tick
    default: break; // Do nothing
  }
  TCNT1 = 0xC2F6; // Reset counter
}
  // Ped switch capture as soon as pressed
ISR (PCINT0_vect) {
  // if green, only then take action
```

```
  if (tick > 5) {
   // Reset timer for yellow
   tick = 0;
   TCNT1 = 0xC2F6; // Restart time count
   PORTB |= B001000; // Turn ON Yellow
  }
 }
 // Main loop
 void loop () {
  // For debug only
  // Serial.println(tick);
  // Optionally you can set sleep mode
 }
```

4.9   Digital Filter

As mentioned earlier, some hardware blocks might be implemented either in hardware or in software. Filter is one of those types of blocks that can be implemented either in hardware or in software. Previously, we studied implementation of filter block in hardware. Here we discuss implementation of filter block in software.

4.9.1   Digital Filter Types

Digital filters are of two types:

1. Finite Impulse Response (FIR),
2. Infinite Impulse Response (IIR).

FIR filter does not have any feedback from output, whereas IIR filter has feedback from output. The generic forms of these filters are:

FIR filter: $y[n] = b0x[n] + b1x[n-1] + \ldots + bNx[n-N]$
IIR filter: $y[n] = 1/a0 \, (b0x[n] + b1x[n-1] + \ldots + bNx[n-N] - a1y[n-1] - a2y[n-2] - \ldots - aQy[n-Q])$

Here $x[n]$ is input signal, $y[n]$ is output signal, and a & b are coefficients for various output and input signal samples, respectively.

4.9.2   Implementation of Digital Filters with Arduino Uno

Here are some example implementation of Digital Filters with Arduino Uno codes.

Example 4.15 How you will design a 3rd order FIR filter with the following equation where the streaming data input is coming from ADC 3:

$$y(n) = (x(n) + x(n-3))/2$$

Solution We will need up to input samples of x[3] data, i.e., x[0], x[1], x[2], x[3]. Afterwards, in the loop code, we will need to shift the samples on each cycle by 1 position as follows:

x[3] = x[2]; // Producing sample delay
x[2] = x[1]; // Producing sample delay
x[1] = x[0]; // Producing sample delay
x[0] = analogRead(3); // New data from ADC 3

Next, we will use the following equation to perform the filter operation:

y = (x[0] + x[3]) >> 1; // Right shift by 1 bit is same as div by 2

Note that here we have used right shift by 1 bit to implement division by 2. This right shift operation (typically 1 clock cycle) is much faster that division operation.

Example 4.16 Design a high pass filter (HPF) with the following IIR equation for Arduino Uno where the streaming input data is coming from Analog input 0.

$$y[n] = \propto \left(y[n-1] + (x[n] - x[n-1]) \right) \text{ where } \propto = \frac{RC}{RC+\Delta t}$$

Solution A code is provided below.

```
const float alpha = 0.5; // controls filter response
double data_filtered[] = {0, 0};
// Output data, i.e. y[0] and y[1]
double data[] = {0, 0};
// Incoming data storage, i.e. x[0] and x[1]
const int n = 1; // IIR depth
// Analog Pin 0; change to where analog data is connected
const int analog_pin = 0;
void setup() {
  Serial.begin(9600); // Initialize serial port
}

void loop() {
  // Retrieve incoming next data
  data[0] = analogRead(analog_pin);
  // High Pass Filter using the above IIR equation
  data_filtered[n] = alpha * (data_filtered[n-1] + data[n] - data[n-1]);
  // Store the previous data in correct index
  data[n-1] = data[n];
  data_filtered[n-1] = data_filtered[n];
  Serial.println(data_filtered[0]); // Print Data
  delay(100); // Wait before next data collection
}
```

Example 4.17 Design a low pass filter (LPF) with the following IIR equation for Arduino Uno where the streaming input data is coming from Analog input 2.

$$y[n] = \propto x[n] + (1 - \propto)y[n-1] \text{ where } \propto = \frac{\Delta t}{RC + \Delta t}$$

Solution A code is provided below.

```
const float alpha = 0.5; // controls filter response
double data_filtered[] = {0, 0};
// Output data, i.e. y[0] and y[1]
double data; // Incoming data storage, i.e. x[0]
const int n = 1; // IIR depth
// Analog Pin 2; change to where analog data is connected
const int analog_pin = 2;
void setup() {
  Serial.begin(9600); // Initialize serial port
}
void loop() {
  // Retrieve incoming next data
  data = analogRead(analog_pin);
  // Low Pass Filter using the above IIR equation
  data_filtered[n] = alpha * data + (1 - alpha) * data_filtered[n-1];
  // Store the last filtered data in data_filtered[n-1]
  data_filtered[n-1] = data_filtered[n];
  Serial.println(data_filtered[n]); // Print Data
  delay(100); // Wait before next data collection
}
```

4.10 Artificial Intelligence

Artificial intelligence (AI) is becoming a major driving force in newer generation ES such as IoTs, wearables, smart devices, and self-driving cars. Here, we briefly discuss a few related aspects to introduce the concept. Any AI system would possess some intelligence or learning ability. Non-AI system are designed by programmers to perform task in a certain way. For instance, an ES street-light controller can be designed by the programmer such that the light turns on when the ambient light falls below certain level by hard-coding a threshold value in the program. We can easily make this system adaptive to deployed setting by using a variable that records the maximum light and minimum light setting at the deployed setting, and set the threshold to a percentage of this range between the maximum and minimum. This system will have some adaptive nature or intelligence.

Example 4.18 Design an Arduino Uno to light up eight LEDs one by one when the light becomes lesser than maximum light. All eight LEDs must be on when the ambient light is lowest and none of the light must be on when the ambient light is maximum.

Solution We can include a variable to keep track of maximum light and another variable to keep track of minimum light. Then we can turn on the LEDs one by one when the ambient light falls below certain percentage. A schematic diagram and a code are provided below.

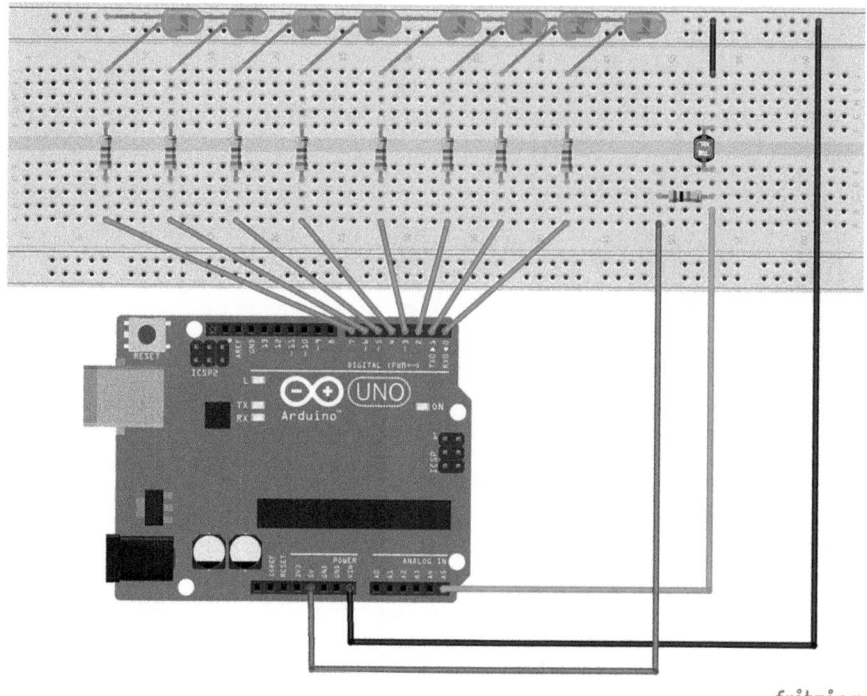

fritzing

```
// Delay 1 is On and 2 is Off
int i = 0;
long randN;
int sensorPin = A0; // select the input pin for LDR
int sensorValue = 0; // to store the value from sensor
float Val = 0.0; // scaled value
float maxVal = 0.0; // maximum recorded value
float minVal = 0.0; // minimum recorded value
// the setup function
void setup() {
// Serial.begin(9600);
//sets serial port for comm, if needed
// initialize 8 digital pins as an output.
 pinMode(0, OUTPUT);
 pinMode(1, OUTPUT);
 pinMode(2, OUTPUT);
 pinMode(3, OUTPUT);
 pinMode(4, OUTPUT);
```

```
  pinMode(5, OUTPUT);
  pinMode(6, OUTPUT);
  pinMode(7, OUTPUT);
}
void loop() {
// read the value from the sensor
sensorValue = analogRead(sensorPin);
//prints the values coming from
// the sensor on the screen
//Serial.println(sensorValue);
// turn all LED off
for (int i = 0; i <= 7; i++) {
  digitalWrite(i, LOW);
}
// Learning process
if (maxVal<sensorValue)
  maxVal = sensorValue;
if (minVal>sensorValue)
  minVal = sensorValue;
// Linear normalized scaling
Val = (float(sensorValue) - minVal)/(maxVal - minVal);
if (Val > 0.1) digitalWrite(0, HIGH);
if (Val > 0.2) digitalWrite(1, HIGH);
if (Val > 0.3) digitalWrite(2, HIGH);
if (Val > 0.4) digitalWrite(3, HIGH);
if (Val > 0.5) digitalWrite(4, HIGH);
if (Val > 0.6) digitalWrite(5, HIGH);
if (Val > 0.7) digitalWrite(6, HIGH);
if (Val > 0.8) digitalWrite(7, HIGH);
delay(100);
}
```

Although this type of intelligent code has some ability to adapt, but this approach is not applicable for complex decision space where decision boundary needs to be non-linear (Fig. 4.5). For those cases, non-linear algorithms are required to achieve high accuracy and sensitivity. Machine learning (ML) is one of the approach that can produce these types of non-linear boundary space without finding the complex process of detailed mathematical modeling. It is becoming one of most used AI technique for ES. The basic concept evolved in attempt to model human learning process. Furthermore, a relative new technique called Deep learning has emerged that is even more powerful, but requires a very large number of training dataset. All of these are subset of AI as shown in Fig. 4.6.

4.10.1 Machine Learning (ML) and Deep Learning

Machine learning (ML) algorithms generate data clusters by finding connections that may not be obvious or expected. ML approach requires a dataset with known true outcome (called Ground Truth). The ML network is first trained with a portion of this

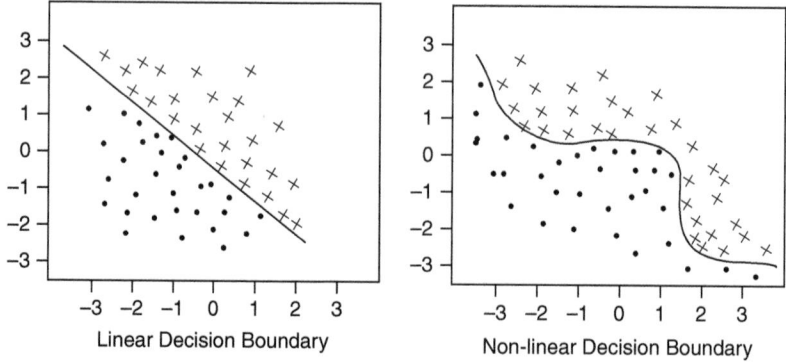

Fig. 4.5 Linear and non-linear decision boundary

Fig. 4.6 Artificial
intelligence, machine
learning, and deep learning

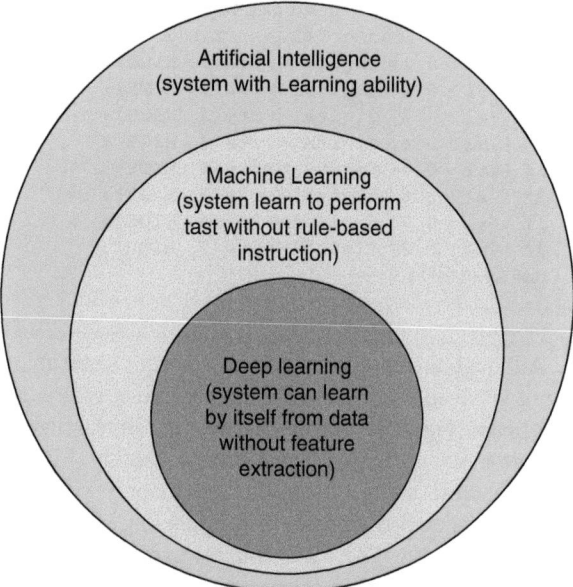

dataset. After algorithm training, the algorithm is tested with rest of the dataset. If the performance (accuracy, sensitivity, specificity, error rate, false positive, false negative, etc.) is satisfactory, then the algorithm is ready for deployment. The scheme is shown in Fig. 4.7.

Artificial Neural Network (ANN) is one of the commonly used network models for ML. It consists of Input layer, Hidden layer, and Output layer. These layer nodes can be fully or partially connected with various weights. When "extracted features" of interest from the training data is fed to the network, the network evaluates the random initial weightages are re-evaluated using a technique called Back-propagation to reduce error. Next iteration uses the newly calculated values. The iterations continue until error falls below an acceptable value. At that point, the weights represent the trained ANN network, which can be implemented for testing and deployment.

Other than ANN, there are many other ML algorithms. Some are supervised and some are unsupervised. Supervised learning ML can be two types: classification type and regression type. Examples of classification type supervised ML are ANN, Support Vector Machine (SVM), Naive Bayes, and Nearest Neighbor. Examples of regression type supervised ML are Linear Regression, Support Vector Regression, Ensemble, and Decision Trees. Unsupervised ML works on Clustering method. Examples include K-means, K-Medoids, Fuzzy C-means, Hierarchical, Gaussian mixture, and Hidden Markov Model (HMM). Some of them are continuous, such as Linear and polynomial regression, Decision Tress, Random Forest, Principal Component Analysis (PCA) and K-means, while other are categorical such as K-nearest neighbor (KNN), Logistic regression, SVM, Naive-Bayes, and HMM.

Classical ML network cannot learn from deployment, rather the network is pretrained prior to deployment. A newer approach to implement ML to be able to learn from deployment is called Reinforcement Learning. This promising approach can adjust ML network parameters based on feedbacks at deployment.

Algorithm Training:

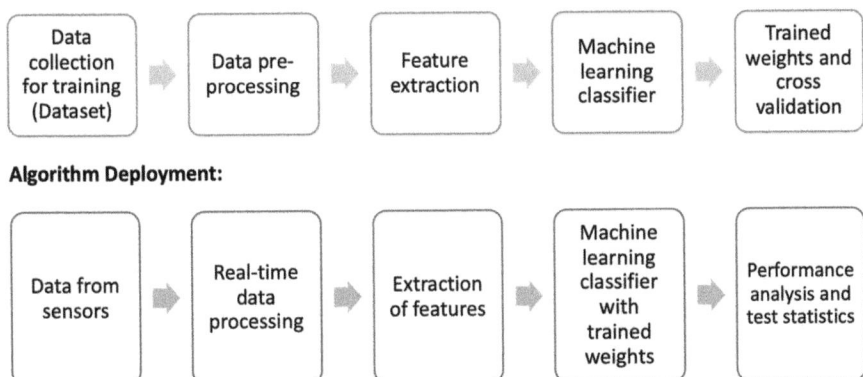

Algorithm Deployment:

Fig. 4.7   Machine learning (ML) algorithm training and testing process

Another very promising AI is Deep Learning. In Deep Learning, the programmer does not need to train the network with extracted features, rather feed the raw data directly to the Deep Learning network. Thus, deep learning network can learn from non-obvious salient features by itself; but it requires a very large amount of data. With the explosion of data (known as Big Data) with large number of IoTs, wearables, and smart devices, deep learning becomes possible and is a prime research interest.

4.10.2   Implementing ANN in Arduino Uno

Implementing ML on OS-less ES is challenging. Currently, approaches are tedious and not suitable. Thus, it is currently recommended to use OS-based ES, such as Raspberry Pi, for ML algorithm implementation. Nonetheless, newer libraries of OS-less platforms are under development and might become suitable in future.

Example 4.19 Write an ANN based machine learning training model for Arduino Uno.

Solution Below is code that implements ANN ML training on Arduino Uno.

```
//Author: Ralph Heymsfeld
//28/06/2018
//With minor modification
#include <math.h>
/***************************************************************
****
 * Network Configuration - customized per network
 ***************************************************************
***/
const int PatternCount = 10; // how many training dataset
const int InputNodes = 3; // feature number
const int HiddenNodes = 5; // larger than input nodes
const int OutputNodes = 2; // output controls
const float LearningRate = 0.3;
const float Momentum = 0.9;
const float InitialWeightMax = 0.5;
const float Success = 0.001; //0.1% error
// training dataset, synthesized
// changed "const char" to "const float"
// e.g. {Peak, Av, std} or {light, humidity, air flow}
// in normalized form
const float Input [PatternCount] [InputNodes] = {
  { 1, 0, 0 }, // 0
  { 1, 0.5, 0 }, // 1
  { 1, 0.2, 0.2 }, // 2
  { 0.5, 0.5, 0 }, // 3
  { 1, 0.5, 0.1 }, // 4
  { 0, 0.5, 0.1 }, // 5
```

```
    { 0.5, 0.4, 0.2 }, // 6
    { 1, 1, 0 }, // 7
    { 0, 0.9, 0.1 }, // 8
    { 0.5, 0.8, 0.2 } // 9
  };
  // ground truth to training dataset
  // {Cooler, Heater }
  // data output is Bool
  const byte Target [PatternCount] [OutputNodes] = {
    { 1, 0 }, // 0
    { 1, 0 }, // 1
    { 1, 0 }, // 2
    { 0, 0 }, // 3
    { 0, 0 }, // 4
    { 0, 0 }, // 5
    { 0, 0 }, // 6
    { 0, 1 }, // 7
    { 0, 1 }, // 8
    { 0, 1 } // 9
  };
  /****************************************************************
****
   * End Network Configuration
   ****************************************************************
***/
  int i, j, p, q, r;
  int ReportEvery1000;
  int RandomizedIndex [PatternCount];
  long TrainingCycle;
  float Rando;
  float Error;
  float Accum;
  float Hidden [HiddenNodes];
  float Output [OutputNodes];
  float HiddenWeights [InputNodes+1] [HiddenNodes];
  float OutputWeights [HiddenNodes+1] [OutputNodes];
  float HiddenDelta [HiddenNodes];
  float OutputDelta [OutputNodes];
  float ChangeHiddenWeights [InputNodes+1] [HiddenNodes];
  float ChangeOutputWeights [HiddenNodes+1] [OutputNodes];
  void setup () {
   Serial.begin (9600);
   randomSeed (analogRead (3));
   ReportEvery1000 = 1;
   for ( p = 0 ; p < PatternCount ; p++ ) {
    RandomizedIndex [p] = p ;
   }
  }
  void loop () {
  /****************************************************************
***
   * Initialize HiddenWeights and ChangeHiddenWeights
   ****************************************************************
***/
    for ( i = 0 ; i < HiddenNodes ; i++ ) {
```

```
   for ( j = 0 ; j <= InputNodes ; j++ ) {
    ChangeHiddenWeights [j] [i] = 0.0 ;
    Rando = float (random (100) ) /100;
    HiddenWeights [j] [i] = 2.0 * ( Rando - 0.5 ) * InitialWeightMax ;
    }
   }

 /****************************************************************
***
  * Initialize OutputWeights and ChangeOutputWeights
  ****************************************************************
***/
   for ( i = 0 ; i < OutputNodes ; i ++ ) {
    for ( j = 0 ; j <= HiddenNodes ; j++ ) {
    ChangeOutputWeights [j] [i] = 0.0 ;
    Rando = float (random (100) ) /100;
    OutputWeights [j] [i] = 2.0 * ( Rando - 0.5 ) * InitialWeightMax ;
    }
   }
   Serial.println ("Initial/Untrained Outputs: ") ;
   toTerminal () ;
 /****************************************************************
***
  * Begin training
  ****************************************************************
***/
   for ( TrainingCycle = 1 ; TrainingCycle < 2147483647 ; TrainingCycle+
+) {
 /****************************************************************
***
  * Randomize order of training patterns
  ****************************************************************
***/
    for ( p = 0 ; p < PatternCount ; p++) {
     q = random (PatternCount) ;
     r = RandomizedIndex [p] ;
     RandomizedIndex [p] = RandomizedIndex [q] ;
     RandomizedIndex [q] = r ;
    }
    Error = 0.0 ;
 /****************************************************************
***
  * Cycle through each training pattern in the randomized order
  ****************************************************************
***/
    for ( q = 0 ; q < PatternCount ; q++ ) {
     p = RandomizedIndex [q] ;
 /****************************************************************
****
  * Compute hidden layer activations
  ****************************************************************
***/
     for ( i = 0 ; i < HiddenNodes ; i++ ) {
```

```
        Accum = HiddenWeights[InputNodes][i] ;
        for( j = 0 ; j < InputNodes ; j++ ) {
         Accum += Input[p][j] * HiddenWeights[j][i] ;
         }
        Hidden[i] = 1.0/(1.0 + exp(-Accum)) ;
        }
   /****************************************************************
****
   * Compute output layer activations and calculate errors
   ****************************************************************
****/
      for( i = 0 ; i < OutputNodes ; i++ ) {
       Accum = OutputWeights[HiddenNodes][i] ;
       for( j = 0 ; j < HiddenNodes ; j++ ) {
        Accum += Hidden[j] * OutputWeights[j][i] ;
        }
       Output[i] = 1.0/(1.0 + exp(-Accum)) ;
       OutputDelta[i] = (Target[p][i] - Output[i]) * Output[i] * (1.0 -
Output[i]) ;
       Error += 0.5 * (Target[p][i] - Output[i]) * (Target[p][i] - Output
[i]) ;
       }
   /****************************************************************
****
   * Backpropagate errors to hidden layer
   ****************************************************************
****/
      for( i = 0 ; i < HiddenNodes ; i++ ) {
       Accum = 0.0 ;
       for( j = 0 ; j < OutputNodes ; j++ ) {
        Accum += OutputWeights[i][j] * OutputDelta[j] ;
        }
       HiddenDelta[i] = Accum * Hidden[i] * (1.0 - Hidden[i]) ;
       }
   /****************************************************************
****
   * Update Inner-->Hidden Weights
   ***************************************************************
***/
      for( i = 0 ; i < HiddenNodes ; i++ ) {
       ChangeHiddenWeights[InputNodes][i] = LearningRate * HiddenDelta
[i] + Momentum * ChangeHiddenWeights[InputNodes][i] ;
       HiddenWeights[InputNodes][i] += ChangeHiddenWeights
[InputNodes][i] ;
       for( j = 0 ; j < InputNodes ; j++ ) {
         ChangeHiddenWeights[j][i] = LearningRate * Input[p][j] *
HiddenDelta[i] + Momentum * ChangeHiddenWeights[j][i];
         HiddenWeights[j][i] += ChangeHiddenWeights[j][i] ;
         }
       }
   /****************************************************************
****
   * Update Hidden-->Output Weights
   ****************************************************************
****/
```

```
      for ( i = 0 ; i < OutputNodes ; i ++ ) {
       ChangeOutputWeights [HiddenNodes] [i] = LearningRate * OutputDelta
 [i] + Momentum * ChangeOutputWeights [HiddenNodes] [i] ;
        OutputWeights [HiddenNodes] [i] += ChangeOutputWeights
 [HiddenNodes] [i] ;
        for ( j = 0 ; j < HiddenNodes ; j++ ) {
          ChangeOutputWeights [j] [i] = LearningRate * Hidden [j] *
 OutputDelta [i] + Momentum * ChangeOutputWeights [j] [i] ;
          OutputWeights [j] [i] += ChangeOutputWeights [j] [i] ;
         }
        }
      }
   /***************************************************************
 ****
    * Every 1000 cycles send data to terminal for display
    ***************************************************************
 ***/
     ReportEvery1000 = ReportEvery1000 - 1;
     if (ReportEvery1000 == 0)
     {
      Serial.println ();
      Serial.println ();
      Serial.print ("TrainingCycle: ");
      Serial.print (TrainingCycle);
      Serial.print (" Error = ");
      Serial.println (Error, 5);
      toTerminal ();
      if (TrainingCycle==1)
      {
       ReportEvery1000 = 999;
      }
      else
      {
       ReportEvery1000 = 1000;
      }
     }
   /***************************************************************
 ****
    * If error rate is less than pre-determined threshold then end
    ***************************************************************
 ***/
     if ( Error < Success ) break ;
     }
     Serial.println ();
     Serial.println ();
     Serial.print ("TrainingCycle: ");
     Serial.print (TrainingCycle);
     Serial.print (" Error = ");
     Serial.println (Error, 5);
     toTerminal ();
     Serial.println ();
     Serial.println ();
     Serial.println ("Training Set Solved! ");
     Serial.println ("-------");
     Serial.println ();
```

```
  Serial.println ();
  ReportEvery1000 = 1;
// pause processor until reset
while(1) {
}
}
void toTerminal()
{
 for ( p = 0 ; p < PatternCount ; p++ ) {
  Serial.println();
  Serial.print (" Training Pattern: ");
  Serial.println (p);
  Serial.print (" Input ");
  for ( i = 0 ; i < InputNodes ; i++ ) {
   Serial.print (Input[p] [i] , DEC);
   Serial.print (" ");
  }
  Serial.print (" Target ");
  for ( i = 0 ; i < OutputNodes ; i++ ) {
   Serial.print (Target[p] [i] , DEC);
   Serial.print (" ");
  }
 /****************************************************************
****
  * Compute hidden layer activations
  ****************************************************************
***/
   for ( i = 0 ; i < HiddenNodes ; i++ ) {
    Accum = HiddenWeights [InputNodes] [i] ;
    for ( j = 0 ; j < InputNodes ; j++ ) {
     Accum += Input [p] [j] * HiddenWeights [j] [i] ;
    }
    Hidden [i] = 1.0/ (1.0 + exp (-Accum)) ;     }
 /****************************************************************
***
  * Compute output layer activations and calculate errors
  ****************************************************************
***/
   for ( i = 0 ; i < OutputNodes ; i++ ) {
    Accum = OutputWeights [HiddenNodes] [i] ;
    for ( j = 0 ; j < HiddenNodes ; j++ ) {
     Accum += Hidden [j] * OutputWeights [j] [i] ;
    }
    Output [i] = 1.0/ (1.0 + exp (-Accum)) ;
   }
   Serial.print (" Output ");
   for ( i = 0 ; i < OutputNodes ; i++ ) {
    Serial.print (Output [i] , 5);
    Serial.print (" ");
   }
  }
 }
```

4.11   Considerations for Memory

As MCU in ES has very limited memory, it is important to consider memory usage for code and data. This is a non-issue for typical computer programming, thus easy to forget in ES coding. A few aspects of memory considerations are discussed here.

4.11.1   Memory Allocations

It is recommended not to use undefined size arrays in ES programming as it might continue to increase during deployment (where ES runs continuously 24/7). Available memory will decrease over time reducing performance of the system. In worst case, the MCU might run out of memory space, and leading to system deadlock. This might need total system reset (use WDT for this purpose), or else not operate at all until manual reset is performed. This will cause disruption of service.

4.11.2   Memory Leak

Another issue to worry about is memory leak. This happens when a program allocates a global memory block for some task; but it does not de-allocate it when existing the program. Next time the program is run, it might consider the previously allocated memory to be unavailable and allocate a new memory block. This is known a memory leak and can reduce performance of the system over time. If this process continues, eventually the MCU will run out of memory and the program will not be able to allocate memory leading to deadlock or improper operation.

Example 4.20 The pseudo-code below has memory lead issue. Identify where, and provide a solution of memory leak.

```
When a button is pressed:
Get some memory to remember the floor number
  Put the floor number into the memory
  Floor:   Are we already on the target floor?
    If so, open the door and wait for passengers
      Nothing else to do, so...
      Finish
    Else:   Wait until the lift is idle
      Go to the (next) required Floor
Release the memory we used to remember the floor number
Finish
```

Solution The memory leak occurs as the IF condition does not free up allocated memory for floor number before "Finish." The code only release allocated memory if the code does not fulfill IF condition. Each time IF is executed, a small memory will lead that was used to remember the floor number. A solution to this memory leak can be as follows.

```
When a button is pressed:
Get some memory to remember the floor number
  Put the floor number into the memory
  Floor:    Are we already on the target floor?
    If so, open the door and wait for passengers
        go to the Cleanup
  Else:    Wait until the lift is idle
      Go to the (next) required Floor
Cleanup:
  Release the memory used to remember the floor number
Finish
```

4.12   Computation and Communication Models of Software

Computation models define components and an execution model for computations for each component. Communication model describes the process for exchange of information between components.

4.12.1   FSM Computation Model

One of the computation models commonly used in ES is Finite State Model (FSM). FSM are models of the behaviors of a system with a limited number of defined conditions or modes, where mode transitions change with circumstances. FSM has four elements: States, Transitions, Rules or Conditions, and Input events. There are two main types of FSM: Moore and Mealy. Moore FSM outputs depend on current state only. Mealy FSM outputs depend on current state and inputs. This model describes various states and data flow where data flow of this model can be flow of data in different computation units of the same system or communication among a distributed system. In a distributed system, models also need to be synchronized.

In case of distributed system or communication with external devices, microcontroller must serve peripheral before the next data appears. This requires synchronization for data transfer without corruption. This can utilize control pins. One simple approach is Baud rate where both transmitter and receiver settle at a predefined data transfer rate. Another approach is blind cycle counting synchronization. Gadfly, Busy-wait, or handshaking mechanism can also be used. Handshaking can be two types: Single handshaking, and Double handshaking.

Asynchronous or non-blocking message passing might be needed if synchronization is not possible (e.g., no control pins or rate not negotiated). In this case, sender does not have to wait until message has arrived. Receiver uses FIFO buffer to receive data. However, buffer overflow might occur if the receiver is slow to retrieve data from FIFO buffer. Alternatively, this can be implemented without buffer, called Rendezvous, where sender will wait sometime for receiver to retrieve data before sending the next data. This is simpler as no buffer required, but reduced performance as the transmission must have some delay between transmissions. Another approach is Extended Rendezvous, where explicit acknowledgment from receiver is required. Sender waits until it receives acknowledgment from receiver of the previous message reception before it transmits the next data.

Synchronization can also be important for internal purpose (within the ES system). This occurs with one master clock. The specification for communication hardware synchronization should be <10 µs, for OS kernel 10 µs to 100 µs, and for application level 500 µs to 5 ms.

4.12.2 Recursion

Recursion is process where a program calls itself. There are three types of recursion: Linear, Tail, and Binary. In order for recursive function to finish, there must a situation where a direct result is generation, called End Condition. It is important to carefully consider termination condition of recursion, otherwise it might lead to race condition or out of memory issue.

4.13   Considerations for ES with OS

For complex systems and processes, OS-based ES is more suitable. There are some additional considerations for this type of system. Some important aspects of this type of ES are discussed in this section.

4.13.1 Thread Scheduler

If multiple programs need to be executed simultaneously, multitasking approach is commonly used by OS. OS uses thread scheduler to manage this. Scheduler type can be Round Robin, Weighted Round Robin, or Priority based. Performance of thread scheduler is measured with utilization of CPU clocks and resources (such as memory), latency for tasks, and bandwidth. Based on thread scheduler operation, threads can be put in Run state where the thread is executing currently, Active state where the thread is ready to run, but waiting it's turn, and Blocked state where the thread is waiting for an external event or availability of needed resource. Each thread has its own stack. Thread control block (TCB) contains the needed information such as stack pointer (SP) value, next or previous links, ID number, sleep counter, blocked pointer, and priority. Thread scheduler controls which thread will be run in sequence, which threads will be blocked, and if there is any thread dependency.

Thread scheduler can use Round Robin Scheduling. One approach is to assign thread times in a Rate Monotonic manner, where each thread requires to be executed are given equal amount of time in round robin fashion. This is a static type scheduling. Priority can be given based on period of time for data arrival, or maximum execution time. For priority-based approach, thread with the highest priority is executed first. There is no synchronization between tasks. For real-time ES, priority can also be assigned to tasks requiring hard real-time operation, then tasks with soft real-time operation, and then the remaining tasks. This minimizes the latency on real-time tasks. One approach to implement this is to assign dollar cost for delay and minimize cost. Another approach is earliest deadline first, which is a dynamic scheduling. Scheduling can also be based on smallest slack time first, which is also dynamic. Slack time is calculated by subtracting work left on a task from time to deadline.

One issue that might result from priority scheduling is that some tasks might never be allocated. This is called "Starvation." One solution for starvation is to assign a weightage of aging. As threads wait for scheduled, aging value will increase and higher aged threads will have increased priority.

4.13.2   Multithreading

Multithreading is a common process in OS. Thread based multiprocessing some-times access shared global variables. This access to global variables by multiple threads might lead to race conditions. To avoid these, mutual exclusion (or Mutex) is required. However, mutex might lead to deadlocks also. Thus, careful consideration of mutex usage and avoiding deadlocks is essential, however might lead to perfor-mance penalties.

4.13.3   Multitasking

OS-based system can allow multitasking by utilizing thread scheduler. All programs, by nature of programming language, has threads. In consideration of thread execu-tion, threads can be foreground or background. In terms of time intervals, threads can be Periodic which executes at regular intervals (e.g., ADC, motor control), Aperiodic which executes without any regular interval, but execution is frequent (e.g., car speed detection from wheel turning), and Sporadic which executes without any regular interval and infrequently (e.g., keypress interrupt, fault, error condition). Multitasking thread scheduler can be pre-emptive or cooperative.

4.13.3.1   Pre-emptive Multitasking:

In this type, OS has the flexibility to shift from process to process by time-slicing approach. Threads are suspended by a periodic interrupt. Intervals for each thread is determined by the OS based on some other factors, such as priority for access to system resources, and intervals are implemented with timer interrupts. OS maintains the context for each process, i.e., registers, memory allocation, etc. The thread scheduler chooses a new thread to run. Code below shows 3 threads, where OS can decide to switch thread wherever it wants (based on timer interrupt).

void Thread1(void){	void Thread2(void){	void Thread3(void){
Initialization1();	Initialization1();	Initialization1();
while(1){	while(1){	while(1){
Task1();	Task2();	Task3();
}	}	}
}	}	}

4.13.3.2   Cooperative (Non-Preemptive) Multitasking

In this type, threads themselves decide when to stop running and allowing other threads to be scheduled. An example is shown below where *OS_Suspend()* indicates only when scheduler is allowed to switch to a different thread.

void Thread1(void){	void Thread2(void){	void Thread3(void){
Initialization1();	Initialization1();	Initialization1();
while(1){	while(1){	while(1){
Task1();	Task2();	Task3();
OS_Suspend();	**OS_Suspend();**	**OS_Suspend();**
}	}	}
}	}	}

4.13.4   Reentrant

A program segment is reentrant if it can be executed by two (or more) threads. As it might use the same variables in registers and stack, it might lead to deadlock. A non-reentrant subroutine can have a section of code with atomic locking which is a vulnerable window or a critical section. However, it is imperative that reentrants must not have atomic write sequence, otherwise deadlock might occur. Still, error can occur when one thread calls the non-reentrant subroutine, which the critical section (if non-atomic) is interrupted by a second thread, and the second thread calls the same subroutine.

4.13.5   Thread Dependency

Threads might have dependency with one or more threads. Dependent thread must be scheduled after the predecessor thread is complete and sequence constraints are met. A dependency graph shows these tasks dependency. If a task $v2$ is dependent directly after the task $v1$, then $v2$ is dependent task and $v1$ is immediate predecessor of $v2$, while $v2$ is an immediate successor of $v1$.

In addition to task dependency, a dependency graph can also show timing dependencies [5]. This additional timing information can specify arrival time, deadline, and other timing information. Dependency graph can also show shared resources such as shared memory or shared bus.

4.14   Shared Memory Issues

Multithread programs can communicate though shared memory. They can be in the form of shared global variables or pointers, mailbox, FIFO queues, or messages. However, shared memory can cause race conditions, producing inconsistent results. To resolve this Mutex, atomic, and semaphores techniques need to be utilized.

4.14.1   Atomic Sections

Critical sections of the program must be made atomic such that sections are not interrupted or resources are locked while the critical section is being executed. In this sections, exclusive access to shared resource (e.g., shared memory) must be guaranteed. As write access can change the value of the shared resource, it is important to have the section atomic. This can be applied to shared variables or I/O ports. However, read access creates two copies of shared information, one is the original in memory and the other is temporary copy in register. They need to be consistent.

To make a critical section atomic, disable all interrupts at the beginning of the section. In addition, lock the scheduler itself so that no other foreground threads can run. However, this might not be possible in many cases where mutex semaphore needs to be used that blocks other threads trying to access the shared resource or information. This will inevitably delay other non-related operations.

4.14.2   Semaphore

Atomic section with disabling interrupt might not be possible for many cases. Semaphores are used to lock resources for this purpose. To protect race-free access to shared memory S, P(S) and V(S) are semaphore operations to lock and unlock the resource, respectively. These allows mutex operation and perform lock on resources. Tasks requiring S might issue these locks. The tasks will implement semaphore like this:

```
        task a {
  ..
P(S) //obtain lock
          .. // critical section
 V(S) //release lock
}
        task b {
  ..
P(S) //obtain lock
          .. // critical section
 V(S) //release lock
}
```

P represents WAIT (from Dutch word proberen). V represents SIGNAL (from Dutch word verhogen). These are executed with *OS_WAIT* and *OS_SIGNAL*, respectively. Semaphores can be Binary type or Counter type. Binary type represents 1 (Free or event occurred), or 0 (Busy or event not occurred). Counter type represents a number corresponding to available copy of resources. Each thread decrements the number representing the number of threads locking the resource. If the counter is 0, then all of the resources are busy. It can be used for space left in FIFO or number of printers available.

Example 4.21 Using pseudo-code, describe Readers-Writers problem with semaphore.

Solution The reader threads (1) Execute ROpen(file), (2) Read information from file, and (3) Execute RClose(file). The writer threads (1) Execute WOpen(file), (2) Read information from file, (3) Write information to file, and (4) Execute WClose(file). Below is a pseudo-core representation of the Readers-Writers problem, where ReadCount = 0 is a number, mutex = 1 is a semaphore, and wrt = 1 is another semaphore.

```
ReadCount, number of Readers that are open
mutex, semaphore controlling access to ReadCount
wrt, semaphore is true if a writer is allowed access
ROpen
  wait(&mutex);
  ReadCount++;
  if(ReadCount==1) wait(&wrt);
  signal(&mutex);
RClose
  wait(&mutex);
  ReadCount--;
  if(ReadCount==0) signal(&wrt);
  signal(&mutex);
```

```
WOpen
 wait(&wrt);
WClose
 signal(&wrt);
```

Semaphores can be implemented with Spin-lock Binary, Spin-lock Counting, Cooperative Spin-lock, or Blocking Semaphore approaches. Blocking semaphore recaptures time lost in spin operation of spin-lock, thus eliminates wasted time running threads that are not doing any useful work. It is implemented with bounded waiting where once thread calls WAIT and is not services, there is only a finite number of threads that will go ahead.

4.14.3 Waiting and Timeouts

Another important items to consider is waiting and timeouts. The waiting should be bounded. Otherwise, timeout should occur. This provides a solution to deadlock detection. It can be implemented with Wait-for-graph or resource allocation graph. This approach works well for single instance resources.

4.15 Software Performance Measure

Some metrics used for software performance measures are:

1. *Breakdown Utilization (BU):* This represents the percentage of resource utilization below which the RTOS can guarantee that all deadlines will be met.
2. *Normalized Mean Response Time (NMRT):* This is the ratio of the "best case" time interval a task becomes ready to execute and then terminates, and the actual CPU time consumed.
3. *Guaranteed ratio (GR):* For dynamic scheduling, the number of tasks whose deadlines can be guaranteed to be met versus the total number of tasks requesting execution.

Exercise

Problem 4.1: Write an Arduino Uno code for Morse Code of "SOS" by blinking the internal LED (connected to Pin 13).

Problem 4.2: Write an Arduino Uno code for Temperature measurement with a thermistor. The analog data from the sensor must be converted to Fahrenheit and Centigrade. Display the temperature readings on an LCD display.

Problem 4.3: What are different queueing mechanisms in ES? Discuss relative advantages and disadvantages. Provide two examples where each of them is more suitable.

Problem 4.4: Describe the 6 sleeping modes of Arduino Uno in order of lower current consumption. Why interrupt is suitable for sleeping modes?

Problem 4.5: What is time service? What module you will need for time service in Arduino Uno?

Problem 4.6: Write a bitmath code to set the digital pin 5 to output mode without changing any other pin modes. Write another code using bitmath to toggle the pin 5 output value.

Problem 4.7: Write a code for a traffic light (using Red, Yellow, Green LEDs) and a pedestrian switch (using a push button). First implement the code with *delay* function, then implement with Timer2 and PCINT interrupt, with lowest current sleep mode.

Problem 4.8: What are two main type of software buffer? Describe relative advantages and disadvantages.

Problem 4.9: Write a C code that utilizes timer/counter hardware of ATmega328 of Arduino Uno, employs extended standby sleep mode, and generates an output pulse of 10 Hz through pin 13.

Problem 4.10. Write a code to implement a IIR LPF for a push button input that turns on an LED.

Problem 4.11: Write a C code that implements ANN machine learning (ML) algorithm to first train with a set of given data in the code, then runs the ANN ML to monitor 3 analog input values (A0–A2) to turn ON/OFF two LEDs (connected to D6–D7).

References

1. E.A. Lee, "Cyber physical systems: design challenges," 2008 11th IEEE International Symposium on Object and Component-Oriented Real-Time Distributed Computing (ISORC), Orlando, FL, 2008, pp. 363–369
2. B. Tabbara, A. Tabbara, A. Sangiovanni-Vincentelli, *Function/Architecture Optimization and Co-Design of Embedded Systems* (Kluwer Academic Publishers, 2000)
3. J. Stankovic, Misconceptions about real-time computing. IEEE Computer **21**, 10 (1988)
4. Atmel ATmega328 datasheet. https://ww1.microchip.com/downloads/en/DeviceDoc/Atmel-7810-Automotive-Microcontrollers-ATmega328P_Datasheet.pdf
5. P. Marwedel, Embedded System Design: Embedded Systems Foundations of Cyber-Physical Systems (2011)

Chapter 5
Prototyping and Verification of ES

Everything should be made as simple as possible, but not
simpler.—Albert Einstein

5.1 Prototyping Phases

Prototyping of ES is the process of developing a testing and verification of various
aspects of the specifications and requirements for hardware or software or both. It
helps to develop an efficient process to streamline the design of ES toward a
commercial product. Prototyping process starts after specifications and requirements
are set and continues until a commercial design is developed. To test various aspects
of the intended design, ES prototyping can be performed at various levels such as
simulation (or emulation), developmental boards, rapid prototyping, and custom
printed circuit board (PCB).

5.1.1 Simulation and Emulation

Some ES systems might be reasonably well modeled with simulation or emulation
tools. A simulator only captures the needed or some aspects of behavior of a target
system, while an emulator captures all component and aspects of the target system.
Sometimes it is difficult or impossible to have a comprehensive emulator, thus
simulator option is more commonly utilized. It is important to bear in mind that
successful test and verification of simulator result does not guarantee proper operation of ES, as aspects that are not modeled might influence significantly in practical
deployments.

© Springer Nature Switzerland AG 2021   167
B. I. Morshed, *Embedded Systems – A Hardware-Software Co-Design Approach*,
https://doi.org/10.1007/978-3-030-66808-2_5

5.1.2 Prototyping with Developmental Boards

This is the most commonly used process of prototyping for ES. Most MCU manufacturers provide some sort of developmental boards with common peripherals typically needed in various ES projects. Arduino Uno is a developmental board that whose design is publicly available, which made this platform an inexpensive and easy to develop prototypes with freely available codes by open-source community. Raspberry Pi and ESP boards are also open source along with some other platforms. Examples of MCU manufacturer developmental boards include LaunchPad from Texas Instruments (TI), PIC and DSPIC developmental boards from Microchip Technology, STM developmental boards from STMicroelectronics, PSoC developmental boards from Cypress Semiconductor, RL developmental boards from Renesas Electronics, DE developmental boards from Altera, and Edison from Intel.

Manufacturers also provide, most of the time free of charge, a software integrated development environment (IDE) suitable for their MCUs. Examples of IDE are MPLab from Microchip Technology, PSoc Creator and Designer from Cypress Semiconductor, Code Warrior from Motorola, AVR Studio from Atmel, and Quartus from Altera. Although programmers might be able to program the MCU without these IDE, however they are recommended software development environment that provides handy additional capabilities such as organizing project files, probing register values, watches and other capabilities for debugging. However, some manufacturer will require use of additional hardware, such as in-circuit debugger (ICD) and background debug module (BDM), to allow programing and debugging. Examples of ICD are MPLAB ICD by Microchip Technology, MiniProg by Cypress Semiconductors, and MSP Debugger by Texas Instruments.

Some developmental boards can be directly connected to computer with USB reducing cost without debugging capabilities, such as Arduino Uno or Raspberry Pi. Some other developmental boards come with on-board debuggers with extended USB capabilities such as USB Blaster by Altera.

Some of the manufacturers also provide daughter boards (or shields) that can simply plug into the developmental boards and allow additional hardware capabilities. Examples of shields are Sparkfun BOB for voltage level interface, Heart Rate monitor shield, Bluetooth shield, and WiFi shield for Arduino Uno.

5.1.3 Rapid Prototyping

Rapid prototyping techniques, such as 3D printer or 3D production system, allow quick prototyping of models and parts for testing. However, these are non-electronic hardware, e.g., PLA, resin, or metal. They can be generated quickly and behaves very similar to the final product. Rapid prototyping might be used for quick package development of the ES, or actuator design, which can be used for DUT testing and verification. It must be taken into account that rapid prototyping may take longer, be more fragile, and consume more power due to its structure, and have other properties that can be accepted in the verification phase, but not exact after commercial prototyping.

5.1.4 Custom PCB

Custom PCB is superior for testing and verification compared to developmental boards as they closely match final product in terms of size and power consumption, as only the needed components of the developmental boards are included in the PCB. It is essential prototyping phase before finalizing ES product design, thus all ES designers must understand and be skilled in PCB design process.

5.2 Testing

Testing includes the application of various stimulus, often in sets of test patterns, to the input of the ES device under test (DUT). The results are observed, recorded, and compared with target specifications. However, ES systems might be complex in nature to develop testing procedures. For instance, an ES control system of a Data Acquisition System (as shown in Fig. 5.1) has interaction with real world environment. Producing real world physical inputs for all possible scenarios are not trivial.

Another consideration is feedback of ES. Open loop control system without any feedback from output to input might be easier to test. Examples of open loop control system are bread toaster machine and traffic light controller. However, control systems can be closed loop as well, where portion of output is fed back to input to compute error from set point. Here, these output values are compared to desired values, and control software generates control commands based upon the differences between estimated and desired values. Example of such feedback control system is PID controller, where a proportional, integral, and derivative values of error is fed to the control algorithm for stable operation. Algorithms to test all various error

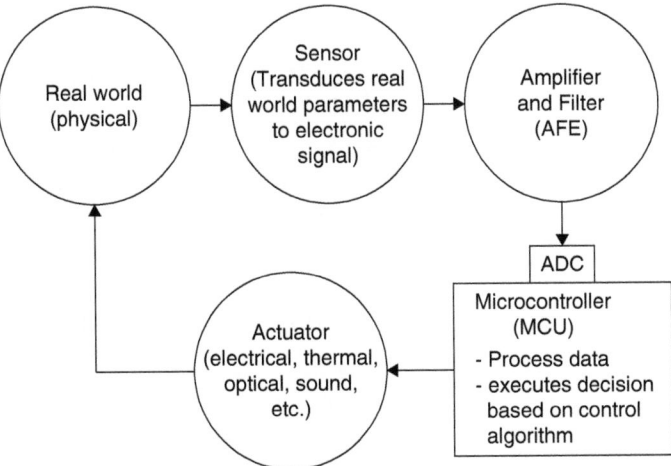

Fig. 5.1 An ES control system for under testing

conditions are not trivial. Further complication can arise from control algorithm in the MCU which might implement non-linear decision space, AI with adaptive algorithm, or reinforcement learning. As non-linear adaptive algorithms can be trained and contains intelligence, test results produced at development phase might not reflect the reality after deployment.

5.2.1   Goals of Testing

The goals for testing are to ensure the specifications and requirements, typically a subset, are met by the prototyped design. This process also helps to develop produces for production test. One of the goals in testing phase is the find out use of test patterns for validating products during production. Tests can be continued to be performed and analyzed after product delivery to customers. Testing is aimed at various levels to conform to the specifications and requirements (Fig. 5.2).

5.2.2   Scopes of Testing

Testing of DUT can be several with each test verifying a subset of specifications and requirements. Scope of each test setting can consist of (a) generation of specific test inputs or test patterns, (b) applying of these test inputs or test patterns to DUT, (c) observation of responses or results which might consist of internal changes and/or output changes, and (d) comparison of these responses or results with expected responses or results.

Fig. 5.2 Various levels of testing for specifications and requirements

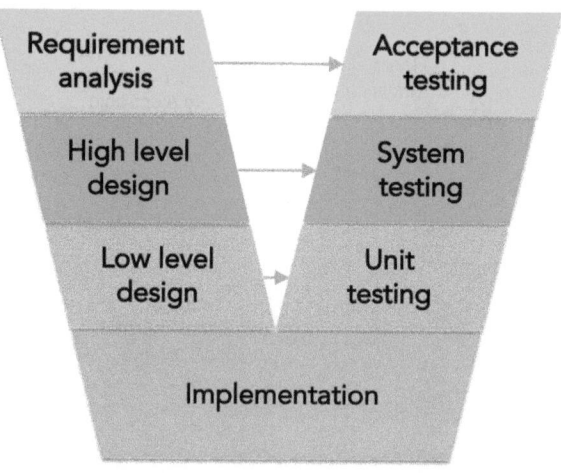

5.2.3   Testing Approaches

Depending on the prototyping platform, testing approach can be three types: (1) Simulation or emulation, (2) Using a developmental board, and (3) Using PCB implementation. The test dataset and observations will be different based on the platform. For instance, inputs for simulation/emulation are computer generated patterns, where inputs for developmental and PCB boards can be computer generator, human generated, or reactive from physical environment.

5.2.4   Considerations of Testing for Various Types of ES

ES testing are important and only functionality verification might not be sufficient for many systems. Follows are some aspects to consider for ES with specific deployment requirements.

1. *Safety-critical ES:* As ES are typically integrated into a physical environment that may be safety-critical. For these cases, expectations for the testing to ensure product quality are higher than for non-safety critical systems. For instance, a fire alarm ES system needs to operate properly all the time as otherwise in case of fire, catastrophic damage can occur.
2. *Time-critical ES:* Testing of timing-critical systems has also need to be validated with the correct timing behavior and constraints such as deadlines. This means that just testing the functional behavior is not sufficient. Example of time-critical ES is the airbag deployment control system of a car. It is not sufficient to have the airbag deployed after crash, it needs to be deployed within milliseconds before the passenger hits any part of the care body.
3. *Cyber-physical systems ES:* Testing of cyber-physical systems, where systems interact with physical systems which are dynamic, needs to be done in their real environment but that may be dangerous or not practical. For instance, testing of an aircraft control unit in mid-air is impractical, rather wind tunnels are utilized for these tests which mimics some aspects of the real environment.

5.2.5   Design for Testability

During the development process of ES product, if testing comes as an afterthought, testing may be very complicated and expensive. Thus, testing considerations must be included early in the design phase. This approach is called design for testability (DFT). This approach is very important for complex systems with FSM and internal states that dictates outcomes.

One solution for DFT is to employ a technique of boundary scan, called JTAG (named after Joint Test Action Group) [1]. It is a serial interface to access complex internal checkout to a set of known inputs. It uses shift registers at inputs to apply the inputs and clock signals to make transition of FSM to execute JTAG functions. Internal checkpoint data are compared with expected results to produce pass/fail report. The system uses 5-wire interface: TDI for test data input, TDO for test data output, TCK for clock signal, TMS for the state of test access port, and TRST for reset signal.

This JTAG technique, developed originally for ICs, can be applied to PCBs as well. It defines a method for setting up a scan chain on a PCB by adding additional circuitry on-board that can apply inputs, change FSMs, and probe internal checkpoint values. Incorporating JTAG within the PCB design process will greatly reduce testing time and streamline the process even after deployment.

5.3   Verification

Prototypes must be validated through testing to progress to next step toward production. Verification can be Functional or Timing.

5.3.1   Functional Verification

Functional verification is defined as the process of verifying that a design meets its specification from a functional perspective. This confirms all functionalities of the DUT that meets or exceeds the specifications and requirements. Depending on the prototyping platform, functional verification aspects will be different. For simulation (or emulation), functional verification typically ensures various levels of abstraction and interaction are acceptable. In case of very high-level abstraction, the verification can be performed quickly, but it hides many details that might not produce accurate results. Lower-level abstraction takes more details into consideration, but they can be slow and tedious. Thus, choosing the level of abstraction is a compromise.

5.3.2   Timing Verification

Timing verification is the process of verifying that a system is fast enough to run without errors at the targeted clock rate to meet the deadlines and other timing constraints. This is complex for OS-based ES as the exact operation delay or scheduling might not be the same at run-time to that at test-time. Thus, timing analysis is done with Worst Case Execution Time (WCET) timing analysis. WCET represents the worst case for a statistically significant large number of tests for each test case.

With regards to timing verification, a set of tasks can be schedulable under a set of constraints. If such schedule exists for that set of tasks and constraints, these tasks are considered schedulable tasks. However, the exact tests are NP-hard in many situations. Thus, sufficient conditions for schedule check might be used to test DUT for "Sufficient tests." The expectation is that small probability of not guaranteeing a schedule will not occur in reality, although it is not improbable. "Necessary tests" include checking of all necessary conditions and show that no other schedule exists that are outside of test set. Although it is difficult, in some cases it might be provable that there is no other schedule that invalidates the timing constraints. For time-triggered system, although some timing constraints can be checked easily, testing of sporadic events and related timing constraints are difficult to validate.

Another type of issue arises for timing verification from various types of scheduling approaches. One of the scheduling methods, earliest due date (EDD) or earliest deadline first (EDF) executes tasks with earliest due date or deadline first. Each time a new task arrives, it is inserted in a queue of tasks sorted by their absolute deadlines. If the newly arrived task is inserted at the head of the queue, then the currently executed task is preempted to pause. This is straightforward approach with sorted list to implement, but uncertainty of incoming task and current queue makes it difficult to test. Other scheduling approaches such as As Soon As Possible (ASAP) where all tasks are scheduled as early as possible might be easier to test, but might have higher failure to meet timing constraints.

5.3.3 Formal Verification

Formal verification is the mathematical process to prove that the system will operate correctly. Formally verified ES are correct by construction. Obtaining this cannot be automated, but requires mathematical rigor. As most ES interacts with stochastic natural processes, it is often not feasible to have a formal verification. If the assumptions of the mathematical model are not met, then the formally verified system can still fail.

5.4 Debugging

Debugging is the process to identify any error or mistakes that might lead to unintended consequences. ES debugging process will be dependent on the prototype platform type. For instance, for simulation or emulation, debugging is performed in software model. Software monitor or profiler can be used to verify the response of the simulator is correct as per specifications and requirements. On the hardware level, such as developmental board and PCB, use of oscilloscope, logic analyzer, ICD, and BDM can be utilized.

5.4.1 Debugging Considerations for Complex ES

Debugging complex systems, such as OS-based ES with multithreading or AI, can have much unpredictability. Scheduling by OS can be based on cost function, which is an algorithm to minimize cost metric for various scenario. Cost function can be maximum latency which is the difference between completion time and deadline. Debugging is challenging in such cases as different run can have different cost values, of which some might be unacceptable.

Furthermore, scheduling also considers priority with hard and soft deadlines. Hard deadlines are critical, which must be met to avoid catastrophe [2]. Periodicity of data and task are also important, as aperiodic data might be difficult to trace. A task is periodic if it is executed once every certain units of time. Examples of periodic data are ADC samples which are set to acquire samples at certain timing intervals. Some tasks can be requesting the processor at unpredictable times. These sporadic tasks are infrequent and difficult to debut. Furthermore, tasks can also be dynamic or online, which further complicates debugging process.

5.4.2 Debugging Difficulty of Time-Triggered ES

Another difficulty in debugging process is to trace time-triggered process. In these ES systems, a temporal control structure of all tasks is established a priori by off-line tool. The temporal control structure can be encoded in a Task Descriptor List (TDL) that contains the schedule and timing information of all tasks. TDL is activated by a synchronized clock tick to perform the action that has been planned for that instance. The debug process might lose timing information or constraint for these DUTs.

5.5 Validation and Evaluation

Validation of ES is the process of checking whether or not a certain design aspect is appropriate and works according to specifications and requirements for its purpose, meets all constraints, and will perform as expected. The result of validation is typically Yes or No. This is done after the ES design is tested and verified for final product. As dependability and correctness are central concerns in ES design, validation is very important and must not be ignored. For instance, a program may have the possibility of deadlock, but nonetheless run correctly for years without the deadlock ever appearing. Due to physical aspects of the ES DUT, not all scenarios are testable as described before, or it might be prohibitively expensive in terms of cost or required time. Thus, ES designers have to be very careful with strong reasoning about the hardware and software so that they are sufficiently validated although it is difficult to ensure there will be no hardware or software errors.

The final step is evaluation of ES, which is the process of computing quantitative information of key characteristics of a certain design aspect at deployment. Design space evaluation (DSE) is based on Pareto-points which informs if the process results a set of Pareto-optimal designs to the user, enabling the user to select the most appropriate design. Some metrics for evaluation of ES designs with multiple objectives are average performance, worst case performance, energy or power consumption, thermal behavior, reliability, electromagnetic compatibility, numeric precision, testability, cost, weight, robustness, usability, extendibility, security, safety, and environmental friendliness. However, performance evaluation is difficult as sufficiently precise estimate is hard to generate due to statistical nature, and a balance must be made between run-time and precision of result. In other words, the performance results are as precise as the input data is. Sometimes average and worst cases performance are computed to resolve this challenge. Worst Case Execution Times (WCET) must be smaller than hard deadlines. If ES maintains WCET < deadlines, then it satisfies the performance of *safeness*. The difference of WCET and deadline is called *tightness*. Evaluation can continue long after the product is commercialized and used by the customers. The long-term target of evaluation is to improve future revisions of the product to improve quality and customer experience.

Exercise

Problem 5.1: Describe testing process of Embedded System Device Under Test. Why it is challenging?

Problem 5.2: What are the differences amount formal verification, timing verification, and formal verification?

Problem 5.3: What are Validation and Evaluation? Why the are important for Embedded Systems?

References

1. IEEE 1149.1 Device Architecture, online: www.jtag.com
2. H. Kopetz, *Real-Time Systems: Design Principles for Distributed Embedded Applications* (Springer, 1997)

Chapter 6
Reliability of ES

Reliability becomes much more important than the cost.—Jeff
Bezos, Founder of Amazon

6.1 Failure, Error, and Fault

A *failure* of ES is an event that occurs when the delivered operation or service of the
ES deviates from the intended correct operation or service [1]. *Error* is the part of the
total state of the system that may lead to its subsequent failure. The hypothesized or
adjudged cause of this error is the *fault*. For an example, assume that a temperature
sensor of a temperature controller ES stops operating, then it is at fault. If this fault
(non-operational sensor) causes the memory variable representing current temperature to be wrong, then this is an error. If this error leads to a situation where the
system is stuck at AC on condition even if the temperature is below the threshold
level, then it is a failure of the ES. Typically, a fault in a system causes an error,
which may lead to a system failure.

6.2 Reliability

Reliability is a statistical probability. Understanding of ES system reliability is
important, as unlike a software code that does not have a degrading reliability over
time, ES reliability decreases over time. This is because all hardware reliability
decreases over time. As ES is hardware-software co-designed system, the overall
reliability is affected if reliability of one is compromised.

To model reliability of ES, we can lend to the concept of hardware reliability
models. For instance, most hardware component failure model follows an exponential distribution. If λ is the coefficient of the probability distribution function, then
the failure density function is $f(t) = \lambda e^{-\lambda t}$ (Fig. 6.1). Here t is the time, when the fist
failure occurs at $t = T$. Thus, the probability of the system being at fault at time t is
given by: $F(t) = \Pr(T \le t) = 1 - e^{-\lambda t}$ (Fig. 6.2). Reliability is the probability that the

© Springer Nature Switzerland AG 2021
B. I. Morshed, *Embedded Systems – A Hardware-Software Co-Design Approach*,
https://doi.org/10.1007/978-3-030-66808-2_6

Fig. 6.1 Failure density function

Fig. 6.2 Failure probability distribution

Fig. 6.3 Reliability probability distribution

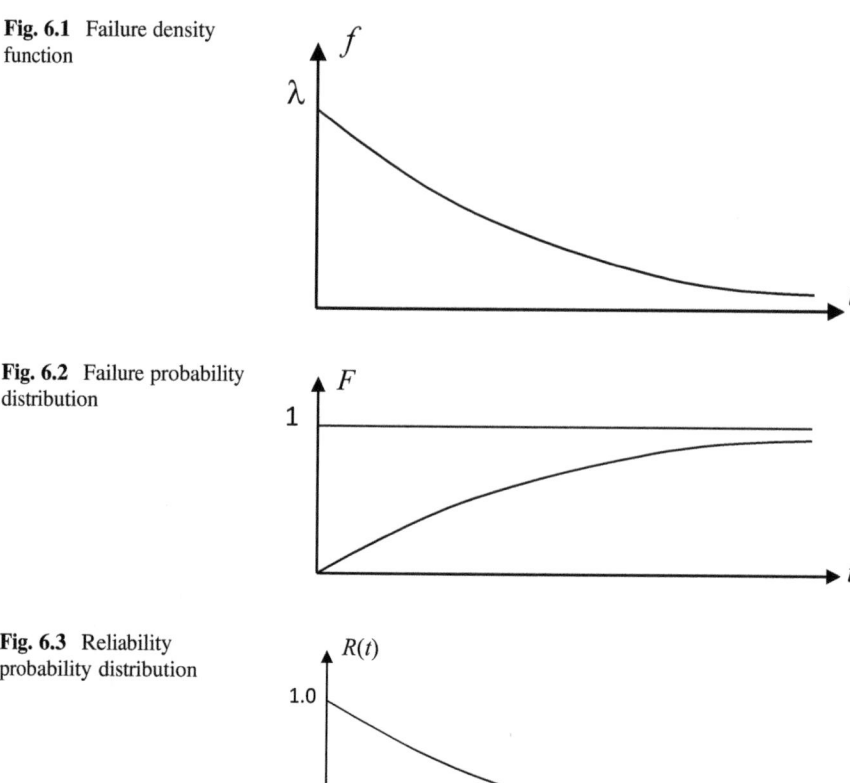

time until the first failure is larger than t. Thus, reliability can be given as: $R(t) = \Pr$ $[T > t]$ for $t > 0, = e^{-\lambda t}$ (Fig. 6.3). In other words, $R(t) = 1 - F(t)$.

The system failure rate at time t is the probability of the system failing between time t and time $t + \Delta t$. This can be determined by taking conditional probability ($\Pr$ $(A|B) = \Pr(AB)/\Pr(B)$) with limit Δt to approach 0. The resulting exponential distribution is $f(t)/R(t) = \lambda$.

The distribution of failure rate is typically similar to "bathtub curve," where failure rate is high at the beginning (called first phase), then becomes very low (called second phase), and increases after a long time (called third phase). For ES, typical expected lifetime before failure, called Failure in Time (FIT), is in the order of 10^9 h.

6.3   Availability of ES

For availability analysis, three metrics are defined as follows:

1. **Mean Time To Failure (MTTF):** MTTF is the time for an ES system to be expected to operate without failure.
2. **Mean Time To Repair (MTTR):** MTTR is the average time for repair of the system.
3. **Mean Time Between Failures (MTBF):** MTBF is the mean time between failures. It is the sum total of MTTF and MTTR.

ES is available during MTTF, but not available during MTTR. Thus, availability can be computed as (Fig. 6.4):

$$\text{Availability} = \text{MTTF/MTBF}$$

Example 6.1 For a repairable ES, the component failure rate is 2 per million hours. If the average repair time is 1000 h, calculate availability of this ES.

Solution   Given, the failure rate λ is $2/10^6$ h.
  MTTF $= 1/\lambda = 500{,}000$ h per failure.
  MTTR $= 1000$ h per failure.
  So, MTBF $= 501{,}000$ h.
  Availability $= 500{,}000/501{,}000 = 99.8\%$.

6.4   Analysis of Fault in ES

For reliable and robust ES design, modes of failures and analysis of fault should be incorporated in design time. For fault analysis, several established methods exist:

- Fault Tree Analysis (FTA)
- Failure Mode and Effect Analysis (FMEA)
- Fault Injection method

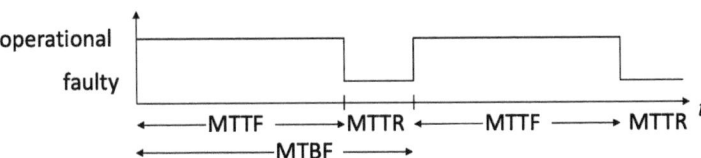

Fig. 6.4 Availability of ES during deployment

6.4.1   Fault Tree Analysis (FTA)

FTA method assumes that for every damage there is a probability and a severity. These damages can result from hazards or risks. To analyze this, FTA uses a top-down method. Analysis starts with possible damage, then tries to come up with possible scenarios that lead to that damage. FTA is represented as an inverted tree diagram. If more than one fault individually can lead to the damage, they are graphed with OR branch connection. If more than one fault must exist simultaneously to lead to the damage, they are graphed with AND branch connection.

6.4.2   Failure Mode and Effect Analysis (FMEA)

FMEA method starts at the component level and tries to estimate their reliability. This is a bottom-up method. The first step is to create a table containing components, possible faults, probability of faults, and consequences on the system behavior. Using this information, the reliability of the system is computed from the reliability of its part.

6.4.3   Fault Injection

Fault Injection method is suitable for ES where fault simulation may be too time-consuming or statistically not feasible due to rarity. In this method, fault is intentionally injected to observe the consequence. Two types of fault injection are used: Internal faults (within the system or local) and External faults (in the environment or external to the system). External faults do not correspond to specification, but it might occur in real-life deployment. For example, one can observe how an ES works under high or low temperature. Fault injection can also be done in software by forcing a wrong data or variable in memory. Advantages of software fault injection are predictability (to reproduce every injected fault in time and space), reachability (can reach storage locations within chips instead of external input/output pins), and less effort (as no test hardware or setup is needed). Fault injection can be also hardware based, although it is a major effort. Examples of hardware (or physical) fault injection are temperature, radiation, electromagnetic interference, and pin connection.

6.5   Fault Testing Approaches

Fault testing needs to consider fault models. Test patterns can be generated and applied to enable a distinction between a faulty ES and a fault-free ES.

6.5.1   Fault Test Pattern

Various algorithms are used to develop test patterns, such as Boolean differences, D-algorithm, and self-test programs. For instance, a stuck-at fault can be simply shorting a pin to supply voltage, ground, or another adjacent pin. This is a hardware fault model, and is a simplification of real situation. The approach allows ES designer to observe how the system behaves due to this stuck-at hardware fault. Similarly, stuck-open fault can be applied by open circuiting a pin or trace. Another way is to introduce delay (thus failing timing constraint) although functionality is not compromised.

6.5.2   Fault Coverage

It might not be possible to test all possible fault scenario. For instance, a certain set of test patterns might not always detect all faults that are possible within a fault model. A metric, fault coverage, is thus defined as: *Fault Coverage = (Number of detectable faults for a given test pattern set)/(Number of faults possible due to the fault model).*

Practical ES products are considered passed when a certain fault coverage values are reached. In most ES, the fault coverage should be in the order of 95–99%, depending on the needs and criticality.

6.6   Dependability Requirements

For safety-critical ES systems, the system as a whole must be more dependable than any of its parts. One way to achieve this is to employ fault-tolerance mechanisms. Typical ES are allowed failures in the order of 1 failure per 10^9 h. However, lower acceptable failure rates can be considered only for ES that are not fully testable and not safety critical. In general, safety must be shown by a combination of testing and reasoning. In the reasoning, abstraction can be used to make the system explainable using a hierarchical set of behavioral models. In addition, design faults and human failures must also be taken into account.

6.7   Reliability with Redundancy

Reliability can either be defined as a characteristic of the device or as a performance measure of it [2]. One approach to improve reliability of a complex ES, Reliability Block Diagram (RBD) technique can be used. Let, A and B are two independent events with probabilities $P(A)$ and $P(B)$ of occurring. Then, the probability $P(AB)$ that both events will occur is the product: $P(AB) = P(A) \cdot P(B)$. However, if two events A and B are mutually exclusive so that when one occurs the other cannot occur, the probability that either A or B will occur is: $P(AB) = P(A) + P(B)$.

According to this technique, for a series of sub-systems of R_1, R_2, R_3, ... R_N, the probability of survival of the system is the probability that all times survive. Thus, the overall probability of these series components R_S can be written as:

$$R_s = R_1.R_2.R_3.\ .\ .\ .\ .R_N$$

$$R_s = \prod_1^N R_i$$

If all sub-system R_i reliabilities are equal (say R), then $R_S = R^N$. Thus, as $R < 1$, higher values of N lead to smaller R_S or overall reliability.

However, if the sub-systems are in parallel, then the overall system fails when all items are failed. This improves reliably. As the probability of individual item failing is $(1 - R_i)$, so the overall probability of failure, $F_P = \prod_1^N (1 - R_i)$. Since $R_P = (1 - F_P)$, we find:

$$R_P = 1 - \prod_1^N (1 - R_i)$$

When all R_i are equal (say R), then the overall probability:

$$R_P = 1 - (1 - R)^{N.}$$

As $R < 1$, so $(1 - R) < 1$. Thus, higher values of N lead to higher R_P or overall reliability.

Example 6.2 Determine the overall reliability of the five sub-systems with reliabilities R_1 to R_5 as connected in the figure.

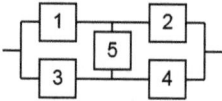

Solution According to Bayes probability theorem,

$$P(B) = P(A).P(B|A) + P(\overline{A}).P(B|\overline{A})$$

Let $P(A) = R_5$, then $P(\overline{A}) = 1 - R_5$. Thus, to solve this complex reliability network, we need to break it down to two different configurations by considering (a) sub-system 5 survives, and (b) sub-system 5 does not survive (i.e. fails).

If (a) sub-system 5 survives, then we can replace sub-system 5 with a short. The network will look like this:

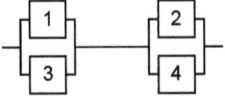

This network has R_1 and R_3 in parallel, and R_2 and R_4 in parallel. Then the overall network is in series. Thus, we can calculate these reliabilities as follows:

$Rs_{13} = 1 - (1 - R_1)(1 - R_3)$
$Rs_{24} = 1 - (1 - R_2)(1 - R_4)$
$Rs_{1234} = (1 - (1 - R_1)(1 - R_3))\,(1 - (1 - R_2)(1 - R_4))$
$Rs^a = R_5\,(1 - (1 - R_1)(1 - R_3))\,(1 - (1 - R_2)(1 - R_4))$

If (b) sub-system 5 does not survive (i.e. fail), then we can replace sub-system 5 with an open. The network will look like this:

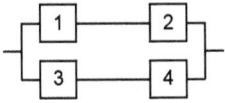

This network has R_1 and R_2 in series and R_3 and R_4 in series. Then the overall network is in parallel. Thus, we can calculate these reliabilities as follows:

$$Rs_{12} = R_1 R_2$$

$$Rs_{34} = R_3 R_4$$

$$Rs_{1234} = 1 - (1 - Rs_{12})(1 - Rs_{34})$$
$$= 1 - (1 - R_1 R_2)\,(1 - R_3 R_4)$$
$$Rs^b = (1 - R_5)\,(1 - (1 - R_1 R_2)\,(1 - R_3 R_4))$$

Then we combine the two sub-networks (a) and (b) using Bayes theorem for overall system reliability:

$$Rs = R_5(1 - (1 - R_1)(1 - R_3))(1 - (1 - R_2)(1 - R_4)) + (1 - R_5)$$
$$\times (1 - (1 - R_1R_2)(1 - R_3R_4))$$

Example 6.3 For the network in the last example, if the reliability values are $R_1 = 0.8$, $R_2 = 0.7$, $R_3 = 0.6$, $R_4 = 0.9$, and $R_5 = 0.5$, calculate overall reliability of the network.

Solution We can use the overall equation calculated in the last example.

$Rs = 0.5 * (1 - (1{-}0.8) * (1 - 0.6)) * (1 - (1 - 0.7) * (1 - 0.9)) + (1 - 0.5)*(1 - (1 - 0.8*0.7) * (1 - 0.6*0.9))$

$Rs = 0.845$

Based on this technique, utilizing two or more sys-systems in parallel can significantly improve overall reliability even if the individual reliability is poor. This approach of redundancy in the ES is defined as existing of two or more means, not necessarily identical, but can accomplish a given single goal or function when needed (without competing with each other). Redundancy can be incorporated in various ways:

1. **Active redundancy:** In this approach, all parallel sub-systems of ES are active all the time to accomplish a given function, even though only one item is required for the function. They can be configured either in pure parallel or in shared parallel form.
2. **Standby redundancy:** In this approach, redundant sub-systems of ES become activated upon failure of the first item. The standby item allows continued operation even after failure of fist item occurred, thus increasing reliability of the overall ES. The standby item can be configured as hot standby (or active standby), cold standby (or passive standby), or warm standby. Hot standby is operating in normal condition even if not needed. Cold standby is not operating normally but becomes available (might lose data or functionality temporarily). Warm standby is normally active but not loaded, thus reduces temporary functional loss but consumes little power when not needed.
3. **R-out-of-n systems:** In this approach, sub-systems of ES consist of n-items in which r of the item must function. Example of this type of redundant ES is RAID configuration of hard disks, which has the highest failure rate in a typical ES.

Exercise

Problem 6.1: What are the differences between failure, error, and fault? Describe with an example.

Problem 6.2: Define MTTF, MTBF, and MTTR. How availability of ES is related to these metrics?

Problem 6.3: For a repairable ES, the failure rate is 5 failures per million hours. If the mean time to repair is 300 h, determine the availability of the system.

Problem 6.4: Describe 3 fault analysis methods for ES.

Problem 6.5: How can you improve reliability of an ES system? Give two examples.

Problem 6.6: For an ES with network of sub-systems shown below with reliabilities $R_1 = 0.1$, $R_2 = 0.7$, $R_3 = 0.2$, $R_4 = 0.8$, and $R_5 = 0.4$, calculate the overall reliability of the ES.

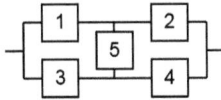

References

1. J.C. Laprie, Dependable Computing and Fault Tolerance: Concepts and Terminology, Proc. 15th IEEE Int'l Symp. Fault-Tolerant Computing (FTCS-15), pp. 2–11, June 1985
2. R. Denning, Applied R&M Manual for Defense Systems, 2012

Chapter 7
Optimization of ES

Genius is one percent inspiration and ninety-nine percent
perspiration.—Thomas Edison, inventor of light bulb

7.1 Optimization Challenges of ES

Evaluating ES typically requires various competing metrics, and optimizing performance of all metrics is not possible. As many different criteria are relevant to ES,
asserting a design as the most optimal is problematic. For instance, ES design criteria
or metrics can be cost, weight, average performance, worst case performance, energy
or power consumption, reliability, thermal behavior, electromagnetic compatibility,
numeric precision, testability, robustness, usability, extendibility, security, safety, or
environmental impact. Although it is often possible to decide that some designs are
"better" than others, it is not often possible to decide the most optimal design.

7.2 Pareto Analysis

Pareto analysis is a well-established multi-objective optimization approach
[1]. Here, each objective must have a defined order or quantitative metrics, where
total order and corresponding order are defined.

7.2.1 ES Solution Space

If X is a m-dimensional solution space for the ES design problem at hand, each
dimension can have its own optimal solutions. Dimensions of solution space can be
processor type, number of processors, clock speed, memory size, code size, bus type,
number of peripherals, etc. The solution space F is derived from n-dimensional
objective space of the Es design problem. Objective space can be cost, speed, power

© Springer Nature Switzerland AG 2021
B. I. Morshed, *Embedded Systems – A Hardware-Software Co-Design Approach*,
https://doi.org/10.1007/978-3-030-66808-2_7

consumption, size, weight, accuracy, reliability, etc. ES design process translates objective space to solution space. ES designers often can choose from a variety of hardware architecture, MCU types, algorithms, communication, and mapping. The optimality of ES thus involves multi-objective optimization problem which has developed theory in Pareto Analysis.

7.2.2 Definitions

The analysis uses some standard terminologies:

1. **Dominate:** Vector u *dominates* vector v iff (if and only if) u is *"better"* than v with respect to one objective and not worse than v with respect to all other objectives.
2. **Indifferent:** Vector u is *indifferent* with respect to vector v iff neither u dominates v nor v dominates u.
3. **Pareto-optimal:** A solution $x \in X$ is called *Pareto-optimal* with respect to X iff there is no solution $y \in X$ such that $u = f(x)$ is dominated by $v = f(y)$. In other words, v is called Pareto-optimal iff v is non-dominated with respect to all solutions in solution space F.
4. **Non-dominated solution:** For a subject of S in solution space, v is called a non-dominated solution with respect to S iff v is not dominated by any element in S.

7.2.3 Pareto Point

A Pareto point is that point with solutions that are not worse than any other solutions. Consider Fig. 7.1 where, for the sake of ease of visualization, only two objectives are

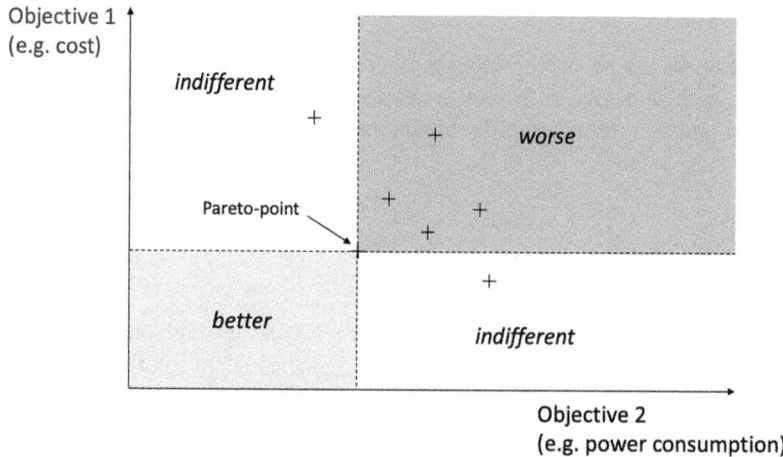

Fig. 7.1 Example of a Pareto point with depiction of better, worse, and indifferent

depicted, with the assumption that minimizing of each objective represents better solution. Example of such objectives and be cost and power consumption.

The Pareto point shown is indifferent to two other solutions, while all other solutions are worse. Thus, the shown solution is Pareto point of all possible design solution of this ES. Pareto points are also referred to as Pareto-optimal points.

7.2.4 Pareto Set

A Pareto set consists of all Pareto-optimal points in the solution space. For any ES solution space, there might be a number of Pareto points. Together they constitute the Pareto set. Figure 7.2 shows an example solution space with indicated Pareto set.

7.2.5 Pareto Front

A Pareto front is the line that includes all solutions of Pareto set in the given solution space. It defines the boundary of optimal ES design possibilities. Figure 7.3 shows an example of Pareto front.

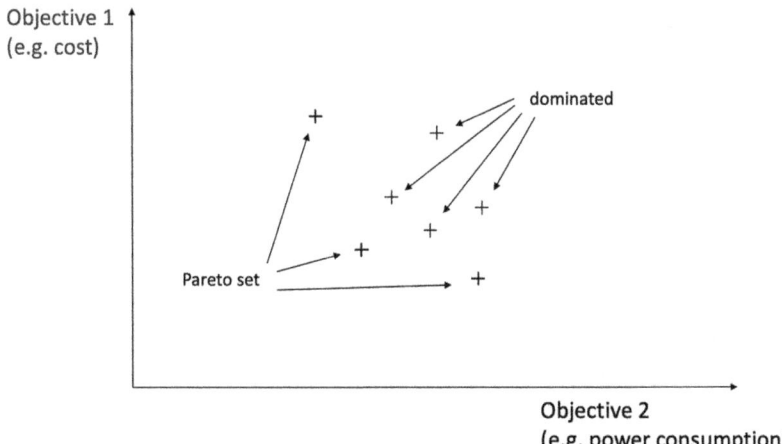

Fig. 7.2 Pareto set in a solution space

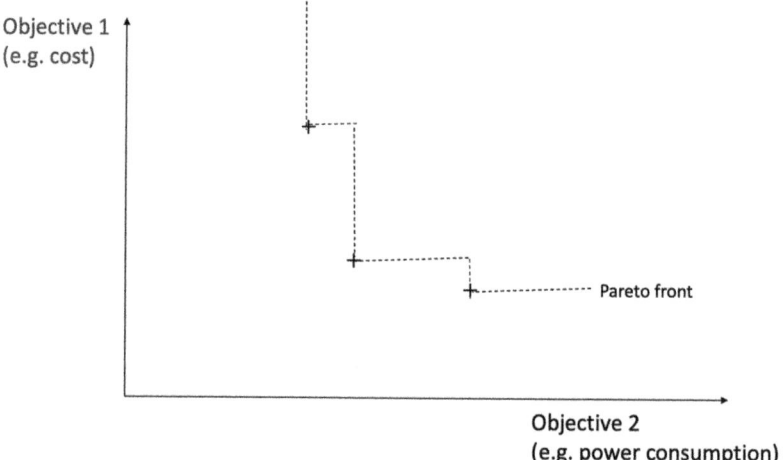

Fig. 7.3 Pareto front in a solution space

7.3 Optimization Scopes

Optimization can be done with various ES parameters depending on the need, application, and criticality. As iterative prototyping for optimization is expensive process, a lot of these optimization process utilizes simulation technique such as Finite element analysis (FEA). However, as simulations only capture some needed aspects of the system, the elimination of errors cannot be guaranteed, rather serves as a probabilistic optimization. Simulation or testing in real environment might be dangerous for some ES. Also capturing all aspects of design in simulation (or emulation) might be too time consuming or resource consuming, that tradeoff must be made.

7.3.1 Power Optimization

Optimization of power is very important for ES. It is especially critical for battery operated ES (such as wearables) and ES that is always on (such as IoTs). Some factors that directly affect power optimization are hours of operation, thermal reduction, higher usability, and more access or availability of ES.

Most MCU allows power optimization through Sleep Modes. Programmers must learn to take advantage of these and make the MCU "*sleep when it can!*" Some other approaches to save power is reduced clock frequency and reduced supply voltage. As higher clock frequency is directly related to higher power consumptions, MCUs allow driving the processor at various clock speeds. While higher clock speed will improve performance and reduce latency, however sometimes these are less important than power consumption. If lower clock of MCU can meet timing deadlines, it is preferred to use lower clock frequency to save power.

Similarly, as lower voltage leads to less power consumption, it is desirable to use letter voltage to operate the hardware. A change of operating voltage from 5 to 3.3 V will save about 56% of power! However, it is important to satisfy voltage related constraints such as lowest operating voltage of all hardware onboard, required voltage levels for interfacing or communication, and the constraint of power supply unit. For instance, LiPo battery produces nominal voltage of 3.7 V; thus supply voltage 3.3 V is desirable to reduce extra hardware need and unnecessary power waste. With regard to battery operated ES, it is also important to consider the use of rechargeable battery (such as LiPo) compared to one time use battery (such as coin cell). Other related considerations are power harvesting or inductive power transfer possibilities. Also note that typically wireless communication modules (such as WiFi or Bluetooth) consume the most power in ES design; thus minimizing use of these during operation is desirable for improved power optimization.

7.3.2 Thermal Optimization

Thermal optimization models are becoming increasingly important for ES designs with complex computation tasks or high payloads. Since temperature becomes more relevant due to increased performance or activities, and temperature has direct effect in usability and dependability of hardware, this optimization needs to be considered in many ES systems. Example of such systems include smartphone, smartwatch, and smart TVs.

The testing and optimization are typically handled with software modeling tools at early stage of design process. These thermal modeling typically uses equivalent electrical models, physics of devices, Joule theorem, heat transfer theorem, and employs well-known techniques for solving electrical network equations or finite element analysis (FEA). Outcomes of these simulations can lead to design parameters for required thermal conductance that reflects the amount of heat transferred through a heat plate or heat sink. The reciprocal of thermal conductance is called thermal resistance. Thermal resistances add up like electrical resistances in equivalent model. Thus, metallic thermal plates or heat sinks can be used that acts as masses storing heat equivalent to capacitors of electric circuit.

7.3.3 Electromagnetic Compatibility Optimization

For some ES designs, the system might be exposed to high amount of electromagnetic interference or emission. For these designs, it is important to perform electromagnetic compatibility (EMC) simulations and validate that EMC properties will not adversely affect ES performance. Examples of such ES systems are car engine controller. EMC simulation is typically performed with FEA tools. The results might dictate need for electromagnetic shields to reduce interference.

7.3.4   Instruction and Memory Optimization

Typically, smaller size of memory leads to lesser energy consumption and lesser access time. Toward this optimization, ES designers should try to employ energy-aware scheduling. In energy-aware scheduling, programmer can change the order of the instructions for optimality which does not change the functionality of the code. The goal of this technique is to reduce the number of signal transitions (i.e., logical state change). This is a popular technique and can be done as a post-pass optimization by the compiler. Most compilers now can perform these optimizations automatically based on the selected optimization level by the programmer in the compiler settings. The compiler optimizes cache operation, pre-fetching, and uses energy-aware instruction selection, which is selecting instructions that minimizes energy consumption from a set of valid instructions. Exploitation of the memory hierarchy can also make a huge difference between the energy consumption of small and large memories.

An example of such optimization is the use of shift operation instead of multiplication or division operation. As shift operations are typically just one clock cycles, it reduces large number of clock cycles if there are many of these operations, such as matrix manipulation. Another possibility is minimizing bit width of leads and stores to fit in a single clock cycle. Most standard compilers can perform these optimizations with energy as a cost function.

7.3.5   Task Level Optimization

Merging and splitting of tasks can lead to more optimized operation of MCU. This requires task level concurrency management in OS-based ES. Granularity, size of tasks, can be optimized to have efficient implementations of required task in a task dependency graph. The most appropriate task graph granularity might depend on context, where task merging and splitting might be required dynamically.

7.3.6   Decimal Number Formats

Fixed-point numbers use decimal at a certain position of the data storage (e.g., Byte). In floating point number format, the data storage is separated in sign bit, exponent, and mantissa. Although floating point number format produces highly precise results, but in a lot of ES applications, precision is not critical, and quantization of fixed-point format might be sufficient for proper operation. In these cases, changing number format from floating point (such as double) to fixed point (such as integer) might lead to a huge performance gain (faster operation) and power savings. One way to use an integer format for fixed point is the multiply the fraction number with a factor at the beginning of operation. For instance, to implement a fourth order IIR filter, a fixed-point format (16-bit) required only 215 clock cycles compared to 2980 clock cycles with floating point format, a saving of about 15 times, while the functionality was maintained [2].

7.3.7 Code Level Optimization

Code level optimization can also improve ES performance. Most compiler currently can perform these optimizations. However, as there is a large variety of instruction sets in different MCUs, the design space exploration for optimized processor can be also explored. One of this technique can be code level optimization for cache performance. Consider the two codes below with the same functionality:

```
for (k=0; k<=m; k++)           for (j=0; j<=n; j++)
  for (j=0; j<=n; j++) )           for (k=0; k<=m; k++)
    p[j][k] = ...                  p[j][k] = ...
```

For the code on the left side that uses inner loop variable as row in the array call performs poorly compared to the code on the right side that uses inner loop variable as column in array call, based on memory architecture of MCU. This is because cache will pull a block of columns of data from main memory in each cache miss operation. For the left side code, there will be a lot of cache misses as the next operation data is further away in memory, leading to a lot of wasted clock cycles. For the right side code, there will be less cache misses as the next operation data are already fetched from main memory to cache.

Another code level optimization is loop fusion. Compare the two codes below:

```
for (j=0; j<=n; j++)           for (j=0; j<=n; j++)
  p[j] = ... ;                   {p[j] = ... ;
for (j=0; j<=n; j++) ,             p[j]=p[j] + ...}
  p[j]=p[j] + ...
```

For the left side code, the loop operation is invoked twice, whereas right side code merges the operations in one loop operation. Thus, the right side code produces better locality and suitable for parallel execution.

Finally, loop-unrolling technique can also be explored. For example,

```
for (j=0; j<=n; j++)           for (j=0; j<=n; j+=2)
  p[j] = ... ;                   {p[j] = ... ; p[j+1] = ...}
```

Here the code on the left hand side has twice the number of loop cycling compared to the code on the right, thus improving performance. This provides better locality for access to p. Also, there will be less branches per execution of the loop for the right had side code, that leaves more opportunities for optimizations. The tradeoff can be between code size and clock cycle improvement. At the extreme case, completely unrolled loop (no branch) can be used; however, code size will increase drastically and might not be practical in a lot of ES MCUs. Thus, as these optimizations are architecture dependent, an architecture aware compiler should be able to select the best code implementation for optimal operation.

Exercise

Problem 7.1: What are the key challenges for optimization of ES designs?

Problem 7.2: What is Pareto-Optimal set? Why this technique is useful for ES designs?

Problem 7.3: How power optimization of ES can be achieved? Provide 3 approaches for power optimization of ES.

Problem 7.4: Explain code level optimization of ES with two examples.

References

1. E.M. Kasprzak, K.E. Lewis, Pareto analysis in multiobjective optimization using the colinearity theorem and scaling method. Struct. Multidiscip. Optim. **22**(3), 208–218 (2001)
2. K.-I. Kum, et al., A Floating-point to Fixed-point C Converter for Fixed-point Digital Signal Processors, *2nd SUIF Workshop*, 1996

Chapter 8
Impacts of ES

> The greatest enemy of knowledge is not ignorance, it is the
> illusion of knowledge.—Stephen Hawking

8.1 Future Trends of ES

Adoption of various ES is growing exponentially worldwide. By 2010, the number of connected devices worldwide surpassed world population [1]. Currently in 2020, there are about 27 billion connected devices, and projection by 2030 is more than 50 billion. Most of these devices are programmed with C, C++, and C# (more than 70%). Assembly and Java are also popular. Most of these ES uses commercial OS (>40%), and then are OS-less ES (30%). Rest are open source or internally developed OS based. ES market has also exceeded $80 B. Most of these markets are in the automotive area, then healthcare electronic devices. Other markets are military, aerospace, telecommunication, consumer electronics, industrial, and others.

8.2 Impact of ES

ES has profound impacts in many aspects of our lives. It has low time to market and low R&D cost. ES provides secure access of information remotely with large data access. Most of them are always on and collects and process these data autonomously, while preserving standards and regulations with innovative edge. ES can assist in local intelligence, remote management, improved performance and sales.

8.2.1 IoTs and Wearables

Internet of Things (IoTs) are one of the most important sectors in ES market. There are about 4 billion people connected through these devices using 25 million apps.

© Springer Nature Switzerland AG 2021
B. I. Morshed, *Embedded Systems – A Hardware-Software Co-Design Approach*,
https://doi.org/10.1007/978-3-030-66808-2_8

These produce about 50 trillion GBs of data. The next big area of ES is wearables that include smart watch, smart bands, heart, EKG, and EEG monitors, gait measurements, step counts, earphones, smart glasses, and music players. Some of them can also assist in medical applications along with smartphones, a sector called mobile health (mHealth). These mHealth apps can monitor users in real-time and generate helpful health data such as exercise, sleep patterns, heart abnormalities, and calorie intakes. Some apps are exploring to extend further by communicating data to physicians or medical records for quick access of longitudinal data.

8.2.2 Self-Driving Cars

Another major area is self-driving car (also known as autonomous vehicles). The enabling technology of this sector is a combination of deep learning and real-time systems along with improvement in sensor technologies, and automated image processing. Currently in 2020, there are no fully self-driving capable cars on the market, but they can be expected in near future. There are six levels of autonomy of cars:

Level 0: These cars do not have any automation. Humans must be in control all the time and responsible for operating the vehicle.

Level 1: These cars provide some level of driving assistance. Examples of assistance can be adaptive cruise control or auto emergency braking.

Level 2: These cars have partial driving automation. The adaptive driver assistants in these cars can take control of car in certain situations by both steering and accelerating or decelerating.

Level 3: These cars also have partial driving automation, but are cognizant of surrounding environment. These can make informed decision by themselves, such as changing lanes and taking exits from highways.

Level 4: These cars will have high driving automations. These cars will not require human interactions except manual override. They will be able to complete full driving from start to finish, including left turns at intersections, understand traffic signals and rules, and acknowledge other vehicles and pedestrians on the road.

Level 5: These will be the highest level of autonomous car, also known as full self-driving car. They will not require any human attention, take dynamic decisions on the road, fully take responsivity from start to finish, and will not require any human intervention. They might not even have traditional drivers control such as steering wheels, gear selectors, or brake pedals.

8.2.3   Electric Vehicles (EVs)

Another complementary, not dependent, ES technology is electric vehicles (EV). These cars operate with electric battery and motor, instead of conventional internal combustion engine (ICE) that uses gasolines. This new car technology made practical with drastic improvement of rechargeable batteries such as LiIon and LiPoly. The high energy density of these batteries along with drastic reduction of related manufacturing cost (about 8 times lower in 9 years). EVs are simpler in construction (lesser moving parts, no transmission), have lower running cost (higher efficiency, lower energy cost, do not require engine or transmission oil services), and operates for decades (dependable battery and motor). Thus, the question is not if EV will take over ICE technology, but when. It is anticipated that within the next 30 years, almost all vehicles on streets will be EVs.

8.2.4   Robots and Humanoids

Another major impact of ES will be adoption and utilization of robots and humanoids in society. Many research and innovative companies are developing technologies. Robotic home vacuum systems are now commercially available. Many other technologies are being tested such as cafeteria servers, hotel receptionists, and bomb diffusers. Technologies being developed by Boston Dynamics or MIT cheetah can be fully humanoids or animal assistants that will be assisting humans in many aspects in future.

8.2.5   Drones and UAVs

Drones and unmanned aerial vehicles (UAVs) also important ES markets. Current use of recreational drones will find more applications and acceptances, along with potential influx of commercial drones for various services such as package delivery, weather forecast, and passenger commute. UAVs also might become more consumer based with applications such as rapid courier services.

8.2.6   Smart Power Grid and Other Technologies

ES are also important for smart power grid controller with connected remote operation as well as smart operation (e.g., load from the grid or to the grid decision). Other technologies include smart home, smart office, and smart city with data from connected technologies and AI based processing of this large data for useful and informative information and knowledge generation.

8.3   Issues and Concerns with ES

While ES has high impact on society and our daily interactions, many issues are anticipated with growing concerns that will need to be addressed. ES designers should make themselves aware of these to be able to consider them during design phases as well as develop solutions for some of these issues and concerns. In general, these can be categories into three types: (1) Regulatory (such as privacy, security, environmental), (2) Ethical (such as moral, social), and (3) Economical (such as transparency, employments, want vs. need).

8.3.1   Regulatory Issues

Technologies enabled by ES such as IoT, self-driving cars, and drones are drastically and dramatically changing human interactions and usages. They are changing too fast for regulatory bodies to evaluate safety and security, and to develop strategies. Issues such as how private data will be handled and managed are still unclear. Another major issue arising from ES with possible human harms such as self-driving cars, robots, or humanoids. It is unclear who will be responsible if the driver is not driving the car, for example. Legal ramification might be tricky. Is the software programmer legally responsible for error in code or bug? If patient dies due to mistake or glitches in mHealth technologies, is the first responder or doctor responsible, or the mHealth technology company? Other regulatory questions that are gaining attention includes environmental sustainability and controlling pollution. Questions regarding how to regulate poisonous material exposure and to regulate radiation exposure are major concerns.

Some more regulatory issues that are concerning are invasion of privacy, data security concerns, hacking of connected devices, vulnerabilities exploited by malicious parties, and environmental pollution due to disposal of broken ES. In these regards, ES designers must make themselves familiar with existing regulatory guidelines:

- **Safety:** International Electrotechnical Commission (IEC 61508), safety integrity level (SIL), IEC 61511 Process industries, Automotive Safety Integrity Level (ASIL) ISO 26262.
- **Aviation:** US Federal Aviation Administration (FAA)—Aerospace (DO-178B and DO-254); DO-178B defines five levels of criticality: Level A—Catastrophic, Level B—Hazardous, Level C—Major, Level D—Minor, Level E—No effect.

- **_Transportation:_** automotive (ISO 26262); National Transportation Safety Board (NTSB); DOT's National Highway Safety Administration (NHTSA).
- **_Wireless radiation:_** CE certification for maximum emission and SAR.
- **_Medical devices:_** US Food and Drug Administration (FDA) regulations—medical (IEC 60601) for device classes [2]:

  - _Class I:_ These devices present minimal potential for harm to the user and are often simpler.
  - _Class II:_ Devices between Class I and III (Most medical devices).
  - _Class III:_ These devices usually sustain or support life, are implanted, or present potential unreasonable risk of illness or injury.
  - _Exempt:_ If a device falls into a generic category of exempted Class I devices.

8.3.2 Ethical Concerns

Human society are founded and prospered with ethical norms and benevolence. Machines, on the other hand, does not need to be bound by these ethical reasoning, rather would benefit from purely greedy (such as game theory). This raises ethical concerns for future ES dependent society where greed might be acceptable rather than benevolence. However, this erases humanity and caring for others, and can change societal norms. Not only interpersonal interactions and relationships will be hurt, this can denigrate to decrease of empath and increase of self-centeredness. These raises ethical questions if greed, worldly goods, and winning at the cost of others are acceptable, or sacrifice and losing for humane sake should be cherished. Philosophical questions arise like _"Is it better to live as a monster or die as a good man?"_ (Shutter Island).

Moral hazards are also might be compromised. Technology bound society might ignore reality and believe in misinformation and disinformation. ES devices tend to empower individuals, but ignores others in surrounding area or in society. Over-reliance on technology, connected devices, smart algorithms, and robotics might breakdown social interactions and sense of belonging within the family, friend, and community. Social isolation is a rising concern observed by psychologists with overly smart device dependent individuals.

A well-known ethical dilemma of self-driving car is as follows. In a situation where an accident is imminent, should the self-driving car save its passenger even if it has to harm people around? In human driven cars related accidents, the decision is made in split second by the driver, but in case of self-driving cars, these decisions need to be made by the company or the programmer well before the incident.

Another major ethical issue arises from the use of robotics in warfare. As robots might not be bound by rules of engagement, the warfare becomes unfair and humanity loses.

8.3.3   *Economical Concerns*

ES will impact economy and workplaces. If robotics replaces human workers, lost jobs might affect the workers families. Displaced workers might need to re-train and continue life-long training to newly created jobs with a different set of requirements and expectations. Another economical issue arises from the fact that gadgets like ES devices are sometimes driven by peer-pressure, fashion, or trend, rather than need. This can lead to waste of money and more importantly, time!

Exercise

Problem 8.1: How ES will impact future society? Explain with three ES trends.

Problem 8.2: What are the three types of issues arising from the next generation ES? Explain each type with specific examples.

References

1. Cisco IBSG, 2011
2. FDA classification for medical devices: https://www.fda.gov/medicaldevices/resourcesforyou/consumers/ucm142523.htm

Appendix A: Brief Tutorials of KiCad Schematic with SPICE

I believe that if you show people the problems and you show them the solutions, they will be moved to act.—Bill Gates, founder of Microsoft

A Simple DC Circuit Simulation

1. Open the KiCad software in your lab computer. *(Note that this procedure is written based on KiCad version 5.1.2 on a Windows 7 64-bit computer. You might notice discrepancies otherwise.)*
2. For SPICE simulation, we will use Eeschema (aka Schematic Layout Editor). Click on this icon. A new blank Eeschema window will open.
3. Click File -> Save Current Sheet As..., browse to your user folder (Instructor will indicate where you can to store this file), enter a File Name (e.g., Lab5.sch), and click Save. Also, periodically save your work as you progress.
4. To put a component in your schematic (e.g., resistor, capacitor, inductor, power supply, switch, etc.), click the Place Symbol icon on the right-hand side toolbar that looks like an op-amp (a triangle with a + and a - symbol).
5. Click in the middle of your schematic. A Choose Symbol window will show up. *Note: It might take a little while for the first time to populate this library.*
6. In the "Filter" field, type "r," and select device->R from the list for resistor. *(Note: Type "l" and select device->L to insert inductor, and type "c" and select device->C to insert capacitor.)*
7. Click OK, and click in the middle of your schematic to place it. Note: You can press shortcut key "r" to rotate the component 90 degrees.
8. You can zoom in or out using the zoom icons on top toolbar. You can also use the mouse scroll bar to zoom in and out. To zoom fit, select the icon of

© Springer Nature Switzerland AG 2021

B. I. Morshed, *Embedded Systems – A Hardware-Software Co-Design Approach*,
https://doi.org/10.1007/978-3-030-66808-2

magnifying glass with a shaded square on its top right, and drag an area around the resistor.

9. You can put value of the component by right clicking on it, select Properties -> Edit Value... (Shortcut key: V)

10. In the field "Text," enter 1k. (Note: Use k for Kilo, M for Mega, G for Giga, m for Milli, u for Micro, n for Nano, p for Pico). Use the "esc" button of the keyboard to get rid of current selection (mouse will change to cursor).

11. Using the same procedure (Steps 4–10), insert another resistor below this first resistor, and put value of 500. *(Note: You do not need to put unit name, Spice uses the standard units based on components, e.g., Ω for resistors.)*

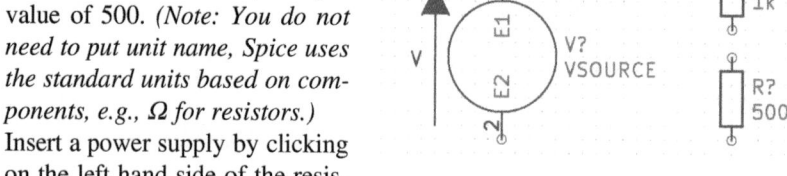

12. Insert a power supply by clicking on the left hand side of the resistors, and type "vsource" in the Filter field, and select "pspice -> VSOURCE."

13. Click OK and place the power supply to the left of the resistors. (as shown beside)

14. To set a voltage, right click on the VSOURCE component, and select Properties -> Edit Value.

15. In the text field, put "10" to apply 10 V DC.

16. To complete the circuit with wires, click green line on the right-side toolbar of Place Wire icon (shortcut key: w). Then click on a starting node for the wire, and click on the ending node for that wire. The software will automatically select the best way to draw the wire. If you want to route manually, click at each corner as you intend. Try to mimic the figure beside to route the wire.

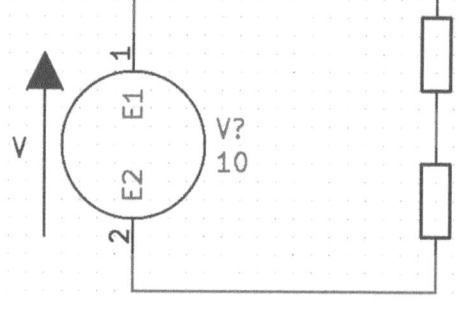

17. For any SPICE simulation, we MUST add a ground (0 node). To do that, click on the icon showing the ground symbol.

18. In the Choose Power Symbol window, type "0" in the Filter, and select "pspice -> 0." Click OK. A triangular ground symbol will be selected.

19. Place this at the bottom of the schematic. Use a wire to connect the 0 node of the ground to the bottom wire of the schematic.

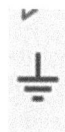

20. Your final schematic show look like this:

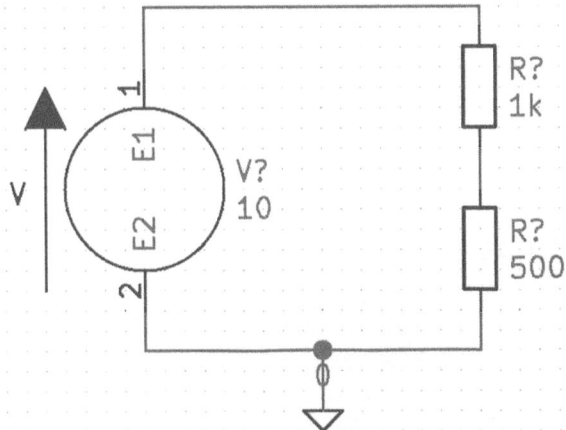

21. After your schematic drawing is complete, the next step is to Annotate Schematic Symbols. On the top toolbar, click the icon with a pen on a paper (Annotate Schematic Symbols).
22. The default selection is sufficient. Just click "Annotate." Then click "Close."
23. You will notice all your components are now numbered (e.g., R1, R2, etc.), i.e., annotated.
24. You can also click the Electrical Rule Check (ERC) button that looks like a ladybug.
25. Click Run. The error list will show if there is any error in the schematic.
26. For our current simulation, it will show an error: Pin 1 (Power input) of component #GND01 is not driven (Net 2). Ignore this error for now.

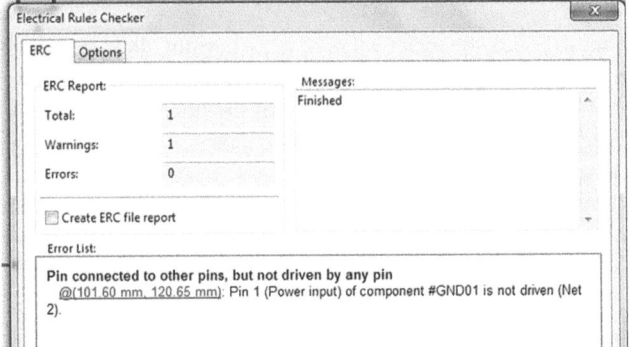

27. Close the ERC window.
28. To simulate the schematic, click Tools -> Simulator. A new simulation window will open.
29. Click on Settings first, then select Transient tab. For this simulation, we will use Time step: 1m, and Final time: 100m. Click OK.
30. Next click Run/Stop Simulation (Green play button).

31. Next click on Add Signals, then in the pop-up menu, select the two voltage signals (hold the Control key to select multiple signals). Click OK.
32. The plot will show two horizontal lines, one at 10 V (supply) and the other at 3.33 V (output of the voltage divider).

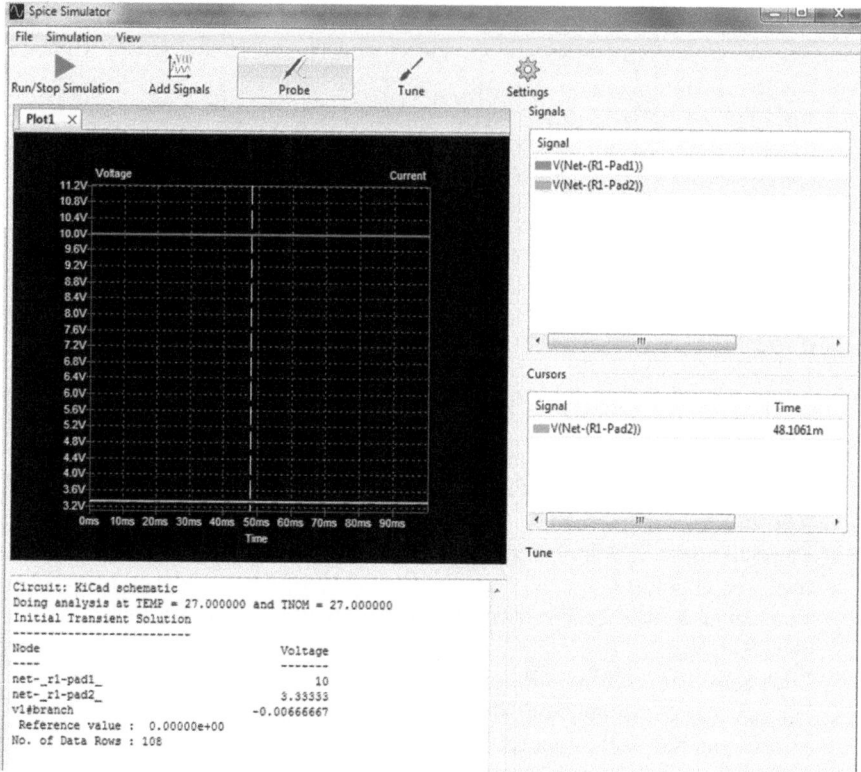

33. If you want to check the exact value of a trace, right click on the corresponding signal in the Signals pane, then select "Show Cursor." You can move the cursor at different time, and see the exact value of the trace in the "Cursors" pane.
34. To observe the current waveforms, click File -> New Plot. Then use Add Signals to plot all three current signals.
35. To zoom in the current waveform, draw a narrow rectangle to zoom the current waveform like this:

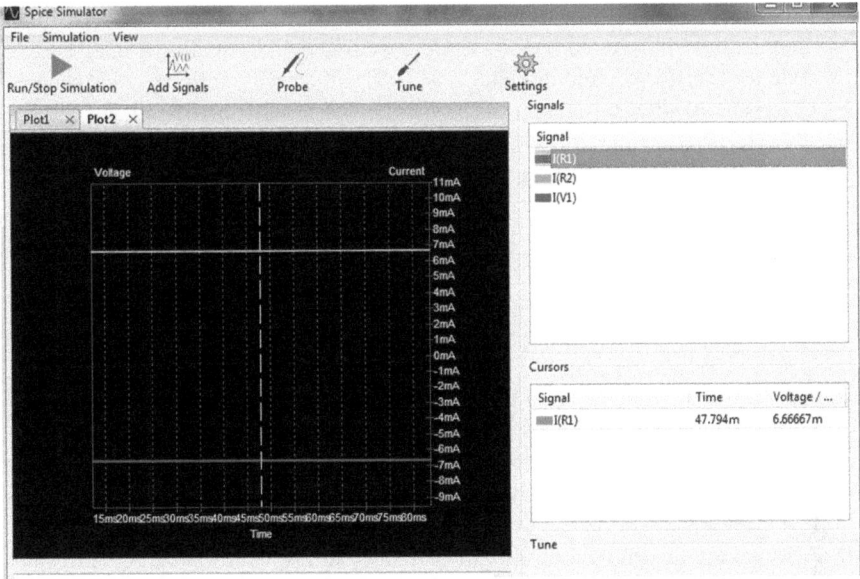

36. Note that I(R1) and I(R2) are exactly the same (overlapped) at 6.67 mA, and these currents are exact opposite of I(V1) at −6.67 mA.

A Simple AC Circuit Simulation

1. Open the KiCad software in your lab computer. *(Note that this procedure is written based on KiCad version 5.1.2 on a Windows 7 64-bit computer. You might notice discrepancies otherwise.)*
2. For SPICE simulation, we will use Eeschema (aka Schematic Layout Editor). Click on this icon. A new blank Eeschema window will open.
3. Click File -> Save Current Sheet As..., browse to your user folder (Instructor will indicate where you can to store this file), enter a File Name (e.g., Lab6.sch), and click Save. Also, periodically save your work as you progress.
4. To put a component in your schematic (e.g., resistor, capacitor, transformer, power supply, etc.), click the Place Symbol icon on the right-hand side toolbar that looks like an op-amp (a triangle with a + and a - symbol).
5. Click in the middle of your schematic. A Choose Symbol window will show up. *Note: It might take a little while for the first time to populate this library.*
6. In the "Filter" field, type "transformer," and select Device-> Transformer_1P_1S from the list.

7. Unfortunately, this built-in transformer symbol does not contain any SPICE model. So we have to add our own. Here we will do it by writing a small SPICE code for transformer sub-circuit.

8. Using a text editor (e.g., Notepad), type the following sub-circuit code.

```
.subckt transformer 1 2 3 4
RP 1 11 50
L1 11 2 2000
RS 4 44 10
L2 44 3 20
K L1 L2 0.99
.ends transformer
```

9. Save this file with name "transmodel.sub" in the same project folder. While saving in Notepad, make sure "Save as type" is selected as "All Files (*.*)." Otherwise Notepad will add .txt extension. To verify that the proper file extension is saved, ensure "Hide File Extension" is turned OFF in the Windows File Explorer.

10. After saving this sub-circuit file, we will need to associate this SPICE model to the transformer symbol on your schematic. To do this, right click on the transformer symbol, and select Properties -> Edit Properties (shortcut key: E).

11. In the Symbol properties window, click Edit Spice Model... button.

12. In the Spice Model Editor window, click the Model tab. Using the Select File button of the Library, browse to the project folder, and select "transmodel.sub." You will need to change file type to "All files (*.*)" in the browsing window to see this .sub file.

13. Select the transformer model, and you will see your sub-circuit model in this window. This transformer model is a step down transformer of 10:1 ratio. Click OK -> OK to confirm. (Note: This is a read-only interface, if you want to make any change, you will need to edit the .sub file with Notepad.)

14. Now we will add an AC source of 120 V. For this, insert the pspice-> VSOURCE symbol. Then right click to select Properties->Edit properties...

15. Click Edit Spice Model, then go to Source tab of the Spice Model Editor.

16. Select the Sinusoidal tab, then set these values: DC offset: 0; Amplitude: 120; and Frequency: 60. Click OK->OK to confirm.

17. On the secondary side, we must add a load resistor for proper simulation. We will use an arbitrary 1 kΩ resistor. Add this resistor on the right side. In the field "Text," enter 1k. (Note: Use k for Kilo, M for Mega, G for Giga, m for Milli, u for Micro, n for Nano, p for Pico).

18. We will also need to place 2 (two) ground (node 0), one for primary side and one for secondary side.

19. Using wire (shortcut key: w), complete the schematic so that it looks like this:

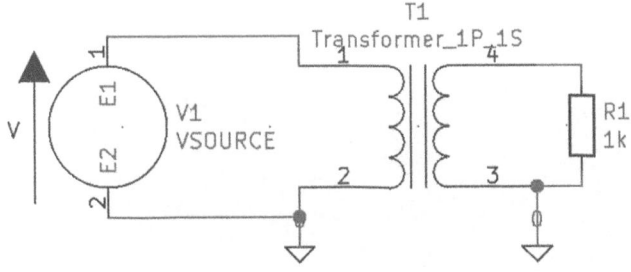

20. Now we will simulate this schematic. Similar to Lab 5, we will perform Annotation, save the schematic, and then open the simulator. We will skip the optional ERC for this lab.
21. After your schematic drawing is complete, the next step is to Annotate Schematic Symbols. On the top toolbar, click the icon with a pen on a paper (Annotate Schematic Symbols).
22. The default selection is sufficient. Just click "Annotate." Then click "Close."
23. You will notice all your components are now numbered (e.g., R1, R2, etc.), i.e., annotated.
24. To simulate the schematic, click Tools -> Simulator. A new simulation window will open.
25. Click on Settings first, then select **Transient** tab. For this simulation, we will use Time step: 1m, and Final time: 50 m. Click OK.
26. Next click Run/Stop Simulation (Green play button).
27. Next click on Add Signals, then in the pop-up menu, select the voltage signals (hold the Control key to select multiple signals). Click OK.
28. You should see this output. If it does not match, re-check your steps and consult with the instructor if needed.

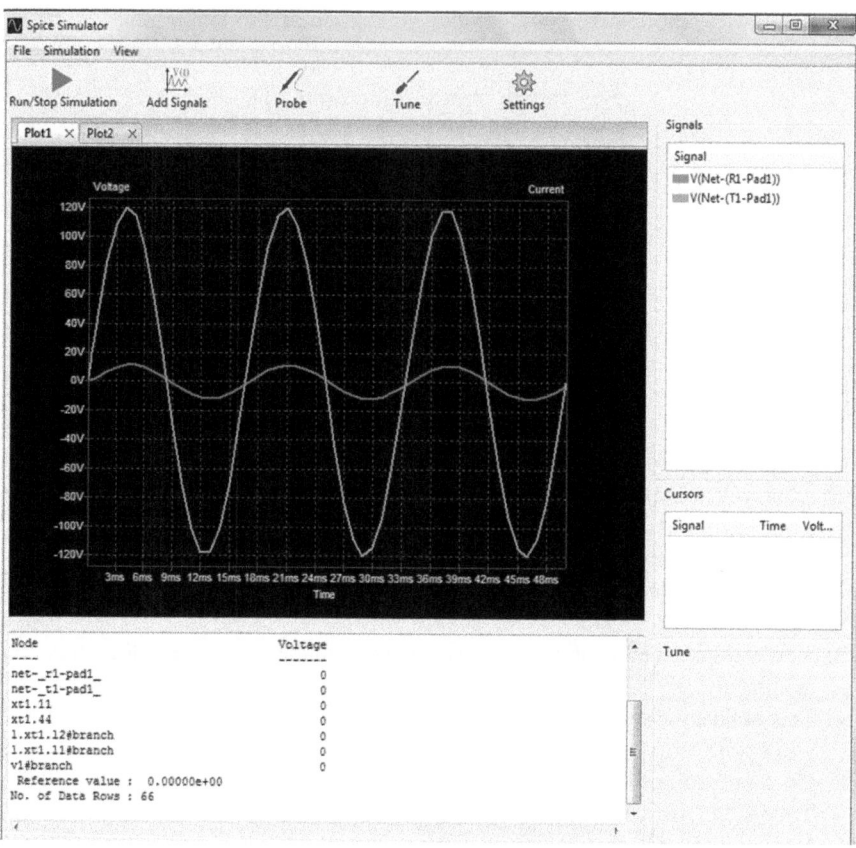

29. To observe the current waveforms, click File -> New Plot. Then use Add Signals to plot all of the current signals.

Appendix B: Arduino Tutorials

For the things we have to learn before we can do them, we learn by doing them.—Aristotle

These tutorials are produced for an Arduino Uno (R3) kit. *All tutorial will require an Arduino board, a computer with Arduino Sketch software and USB port, and a USB programming cable for Arduino.* You can download Arduino Sketch for free from www.arduino.cc website. These tutorials are tested with software version "ARDUINO 1.8.13." For all tutorials, there are some common steps to ensure the board is properly connected/setup. The steps are as follows:

1. Connect the Arduino with the USB cable to the computer.
2. Open Arduino Sketch in the computer.
3. Go to Tools > Board, ensure the board name is correct (e.g., Arduino Uno).
4. Go to Tools > Port. Select the COM port that indicates Arduino Uno.
5. Click the → to download the "blank" code on the board to ensure the board is setup properly. If you get any error message, you will need to resolve it before you can try any of the tutorials.

Blinking LED Tutorial

Objective: Make an LED to blink ON and OFF.

Components: (1) A 2-pin LED (any color), (2) One 220 Ω resistor, (3) A breadboard, (4) Some jumper wires.

Procedure: This tutorial has two parts. In the first part, we will use the on-board LED as our output target. On the second part, we will add an external LED as well.

1. After confirming the board setup is correct, go to File > Examples > 01. Basic > Blink. A new window with C code will pop-up. The main part of the code will look like this:

© Springer Nature Switzerland AG 2021

B. I. Morshed, *Embedded Systems – A Hardware-Software Co-Design Approach*, https://doi.org/10.1007/978-3-030-66808-2

```
void setup() {
  pinMode(LED_BUILTIN, OUTPUT);
}
void loop() {
  digitalWrite(LED_BUILTIN, HIGH);
  delay(1000);
  digitalWrite(LED_BUILTIN, LOW);
  delay(1000);
}
```

2. Click the → to download the code on the board and run the code.
3. An LED beside Digital 13 will start to blink. It should be ON for 1 s, and OFF for 1 s.
4. Click File > Save As. Browse to a folder where you want to save the file. Give a name that is easy to recognize.
5. Scroll through the code to go to "delay(1000)" lines. These numeric values (e.g., 1000) are given in terms of ms. Thus the delay statement produces 1 s delay. Now, change both delay values to "delay(2000)."
6. Click the → to download the code on the board and run the code. Has the blinking rate changed? What is the new rate?
7. Change the delay values to 100 for both. Click the → to download. What do you observe?
8. Change the delay values to 10 for both. Click the → to download. Do you observe something different? Can you explain your observation?
9. In this next part, we will connect and external LED. Change the delay values to 500 for both. Click the → to download. Ensure you are observing the blink rate accordingly.
10. Disconnect the USB cable from the computer. Any electronic circuit connection must be done with power disconnected.
11. On a breadboard, connect an LED and a resistor with jumper wires as shown in the figure below.

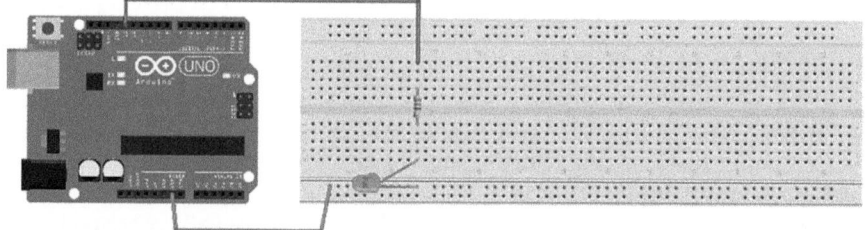

fritzing

a. **Note** that LED has polarity of terminals, i.e., one of the pins (longer) is for positive terminal, and the other pin (shorter) is for negative terminal. This must match for the LED to be ON.
b. **Note** that LEDs must not be connected directly without the resistor. Resistor is required to limit current through the LED, otherwise it might burn out or

reduce its lifetime. Commonly used resistor values are between 200 and 500 Ω for voltages of 3.3–5 V.

12. The positive terminal of the LED connects to the resistor, which connects to Digital 13 of the Arduino board with a jumper wire. Another jumper wire from the negative side of the LED connects to one of the GND terminal of the Arduino board.

13. Re-connect the USB. The code should be running (as Arduino memory is non-volatile).

14. You can change the delay to 200 ms, and download the code to observe the changes.

15. Now, we will change the output port pin to Digital 12. To do this, find the line in the code that says "pinMode(LED_BUILTIN, OUTPUT);". The LED_BUILTIN value is 13, thus this code makes digital port 13 as the output. To change the output to a different port such as Digital 12, we will need to modify this line as follows "pinMode(12, OUTPUT);".

16. We also need to make this new pin change to the digitalWrite statements "digitalWrite(12, HIGH);" and "digitalWrite(12, LOW);".

17. Click the → to download the code on the board and run the code. What do you observe?

18. The reason we are not observing LEDs to blink is that the jumper wire is connected to Digital 13, instead of Digital 12. To change this, first disconnect the USB cable, then change the red wire to Digital 12.

19. Now re-connect the USB cable. You should see the external LED is blinking, but the built-in LED is not. This is because the built-in LED is hard-wired internally to Digital 13, and cannot be changed. On the other hand, external LED can be connected and programmed to any Digital connector pin.

Expected output/observation: The LED should be blinking ON and OFF with specified time delay.

Additional explorations:

(a) Can you make the LED to be ON for a shorter duration (e.g., 0.1 s) and OFF for a longer duration (e.g., 0.5 s)?

(b) Can you connect 5 LEDs of different colors to 5 different digital ports (each LED with a series resistor) and write a code to blink them at different rates?

Analog Light Sensor Tutorial

Objective: Observe analog data from a light sensor using Arduino sketch built-in serial terminal tools.

Components: (1) A light sensor (2-pin analog type), (2) A 10 kΩ resistor, (3) A breadboard, (4) Some jumper wires, (5) A 220 Ω resistor, (6) A 2-pin (white) LED.

Procedure:

1. After ensuring the Arduino is setup/connected, open a new Arduino Sketch window, and type the following code.

```
const int analogInPin = A0;
int sensorValue = 0;
void setup() {
  Serial.begin(9600);
}
void loop() {
  sensorValue = analogRead(analogInPin);
  Serial.print("Analog sensor value = ");
  Serial.println(sensorValue);
  delay(100);
}
```

2. Save the file to your preferred folder with a suitable name.
3. Click the → to download the code on the board and run the code. There should not be any error. If any error, check your code to fix any issue.
4. To prepare the circuit, disconnect the USB cable. The circuit consists of a light sensor (2 pin) in series with a resistor (10 kΩ). None of these has polarity. Use a jumper wires from one side of the light sensor to +5 V, another jumper wise from the other side of the resistor to GND, and a third jumper wire from the middle point to Analog A0. A connection drawing is shown below.

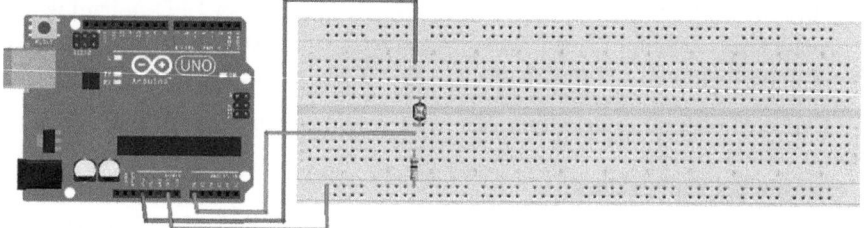

5. Re-connect the USB cable. Click the → to download the code on the board and run the code.
6. To observe the serial port data, go to Tools > Serial Monitor. A new window will pop-up with scrolling data shown after the text: "Analog sensor value =." Note the use of a fixed text and a line break in the code.
7. Note the range of values being displayed. Now using your finger/hand, cover the light sensor so that sensor goes into shadow. The values should decrease significantly. You can also use a piece of paper or something else to cover up the sensor to see the range of values.
8. Cross off (close) the Serial Monitor. Go to Tools > Serial Plotter. A new window will open. (Note that Serial Monitor and Serial Plotter cannot be open at the same time.)

9. Put the light sensor in shade and observe the change of data. Note down the approximate value when it is in full shade (low value) and not in shade (high value). The difference is the range of your light sensor for the given lighting condition (and your circuit setup).

10. Determine the mid-point value (let's say X). Now we will connect an LED to turn ON when the light sensor value is lower than X.

11. Disconnect the USB cable. Update your circuit as shown below, by connecting a series resistor (220 Ω) and an LED (preferably 2-pin White color) connected to Digital 7. Note the polarity of LED.

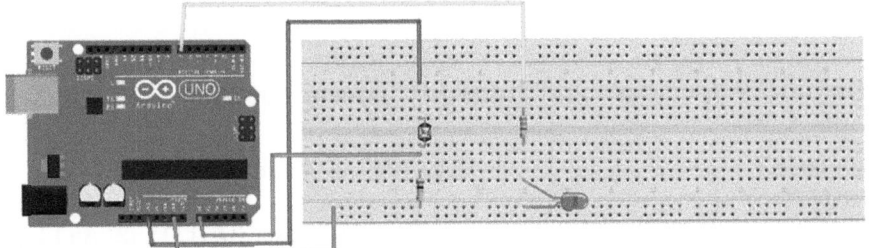

fritzing

12. Modify the code as follows. Here "X" value of 550 is used. If you have a different value, change it accordingly.

```
const int analogInPin = A0;
int sensorValue = 0;
void setup() {
 Serial.begin(9600);
 pinMode(7, OUTPUT);
}

void loop() {
 sensorValue = analogRead(analogInPin);
 Serial.print("Analog sensor value = ");
 Serial.println(sensorValue);
 if (sensorValue < 550) {
   digitalWrite(7, HIGH);
 } else {
   digitalWrite(7, LOW);
 }
 delay(100);
}
```

13. Re-connect the USB cable. Click the → to download the code on the board and run the code.

14. Now when you put shadow on the light sensor, the LED will turn ON. You have just made your night light prototype!

Expected output/observation: The serial monitor will show the streaming data in real-time, and the serial plotter will plot these data. The LED will turn on when there is enough shadow on the light sensor.

Additional explorations:

(a) Can you change to code to turn on the light sensor only when it is sufficiently dark (instead of turning it ON at the mid-range of light sensor)? How you will conduct this experiment to determine X?

(b) Can you change the light to turn on slowly as it becomes darker? This cannot be done with a simple threshold given in this code. What option can you think of?

Digital Temperature and Humidity Sensor Tutorial

Objective: Observe digital sensor data from a temperature/humidity sensor (DHT11).

Components: (1) A digital temperature/humidity sensor (DHT11), (2) An LCD panel, (3) A 10 kΩ potentiometer (pot), (4) Some jumper wires.

Procedure:

1. After ensuring the Arduino is setup/connected, open a new sketch window, and copy the code below. To run this, you will need to download DHT11 library (e.g., adidax-dht11-b7fbbcd.zip) from Arduino site (for dht11.h file). After downloading the zip file, move it to a folder of your choice, then go to Sketch > Include library > Add .zip library... Browse to your folder of choice, and select the DHT11 library zip file.

```
#include <dht11.h>
#define DHT11PIN 8
dht11 DHT11;
void setup()
{
  Serial.begin(9600);
}
void loop() {
  Serial.println();
  int chk = DHT11.read(DHT11PIN);
  Serial.print("Humidity (%): ");
  Serial.println((float)DHT11.humidity, 2);
  Serial.print("Temperature (C): ");
  Serial.println((float)DHT11.temperature, 2);
  delay(1000);
}
```

2. Prepare the circuit with DHT11 (with 3 pin connections: S-Signal, 5 V VDD, 0 V GND) as shown in the figure below. The left side pin marked with S will be connected to Digital 8. The middle pin will connect to 5 V (VDD) and the right-side pin will connect to GND of the Power terminal. Recheck your connection to ensure the wiring is correct before powering on.

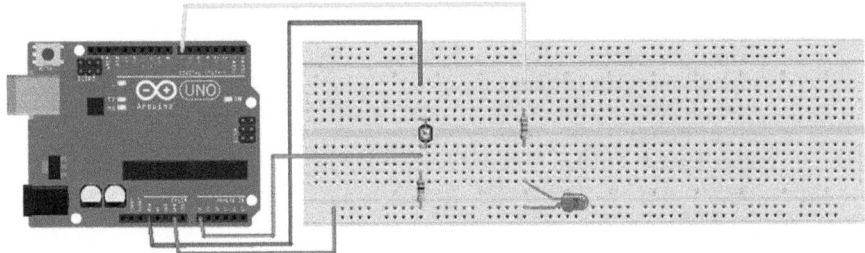

fritzing

3. After connection is complete, connect USB cable and download the code.
4. Open the Serial Monitor. You should see the humidity value and temperature value are shown.
5. First, we will convert the temperature from Celsius unit to Fahrenheit unit. To do this, we need to declare a new float by adding this statement before the setup code:

```
float tempF = 0.0;
```

Then, add this equation before the temperature is printed:

```
tempF = ((float)DHT11.temperature * 9.0 / 5.0) + 32.0;
```

Then, update the serial print statements for temperature display for correct information:

```
Serial.print("Temperature (F): ");
Serial.println(tempF, 2);
```

6. Now, re-download the code and observe the data. It should show the values in Fahrenheit units.
7. To test the sensor working, you can pinch the sensor with fingers to raise temperature slightly or blow into it to change humidity.
8. Finally, we want to display this data in an LCD panel. Connection to LCD panel requires a lot of wiring.
9. The wiring connection table is listed below. Note the LCD panel pin numbers and connect them to Arduino boards. Note some pins are not connected.

LCD	1	2	3	4	5	6	7	8	9	10	11	12	13	14	15	16
Arduino	GND	5V	Contrast	12	GND	11	-	-	-	-	5	4	3	2	5V	GND

For the contrast, you should connect a Potentiometer (Pot) that will allow you to adjust the contrast by turning the pot. To do this, connect two end terminals of the Pot to +5V and GND. Make sure the two fixed terminals (physically on one side of the pot) are connected to VDD and GND. Then, connect the middle terminal (on the other side) to Pin 3 of the LCD module.

10. Update the code as follows:

```
#include <dht11.h>
#include <LiquidCrystal.h>
#define DHT11PIN 8
dht11 DHT11;
float tempF = 0.0;
LiquidCrystal lcd(12,11,5,4,3,2);
void setup()
{
  lcd.begin(16, 2);
  lcd.print("Initializing...");
  Serial.begin(9600);
  delay(3000);
}
void loop()
{
  lcd.clear();
  int chk = DHT11.read(DHT11PIN);
  tempF = ((float)DHT11.temperature*9.0/5.0) + 32.0;
  lcd.print("Humidity:");
  lcd.print((float)DHT11.humidity);
  lcd.print("%");
  lcd.setCursor(0,1);
  lcd.print("Temp.:");
  lcd.print(tempF);
  lcd.print("F");
  delay(1000);
}
```

11. After ensuring all the wiring connection is correct, re-connect the USB cable. Click the → to download the code on the board and run the code.

12. If the LCD is not displaying any text, recheck your wiring. If the LCD backlight is not on, check all your VDD and GND connections. If backlight is on, but you do not see any text, the Pot might be not connected properly. The 3 terminal Pot is hard to connect. Ensure it is all the way in. Also turn the Pot all the way to both sides to see if you see the text. Also check the two fixed terminals (on one side of the pot) are connected to VDD and GND properly.

Expected output/observation: The LCD should display the Humidity value on the first line, and Temperature value on the second line. It will update every second.

Additional explorations:

(a) Can you save and show more information such as maximum and minimum readings of humidity and temperature? Note that there is a physical reset button on the Arduino board beside the USB connector.

(b) As LCD has limited number of character space, what you can do if the text is larger? Can you modify the code to show scrolling display for larger text (e.g., the complete word "Temperature:" and/or Current, Maximum, and Minimum values)?

Remote-Controlled Servo Motor Tutorial

Objective: Control movement of a servo motor with an IR remote controller.

Components: (1) An IR remote control unit, (2) An IR receiver, (3) A servo motor.

Procedure:

1. We will divide the target in two portions. In the first portion, we will test a code for IR remote control with the Serial Monitor. In the second portion, we will test a servo motor movement. Then finally, we will combine the codes from two sections and complete the objective.

2. For the first portion, after ensuring the Arduino is setup/connected, open a new sketch window, and copy the code below. You will need to download IRremote library (e.g., IRremote-2.6.1.zip) from Arduino site (for IRremote.h file). After downloading the zip file, move it to a folder of your choice, then go to Sketch > Include library > Add .zip library... Browse to your folder of choice, and select the IRremote library zip file.

```
#include <IRremote.h>
int RECV_PIN = 2;
IRrecv irrecv(RECV_PIN);
decode_results results;
int value;
void setup()
{
  Serial.begin(9600);
  irrecv.enableIRIn();
}
void loop()
{
if (irrecv.decode(&results))// 0 if no data, or 1
{
value = results.value;// Store results
Serial.println(" ");
Serial.print("Code: ");
Serial.println(value,HEX); //prints the value
Serial.println(" ");
irrecv.resume(); // Ready to receive the next value
}
}
```

3. The IR receiver has 3 pins: GND (G), VDD (R), and Signal (Y). Attach the IR receiver on the breadboard so that the IR sensor is exposed for you to use the remote. The Signal pin connects to Digital 2 (see code).
4. Connect the wiring with the Arduino as shown below:

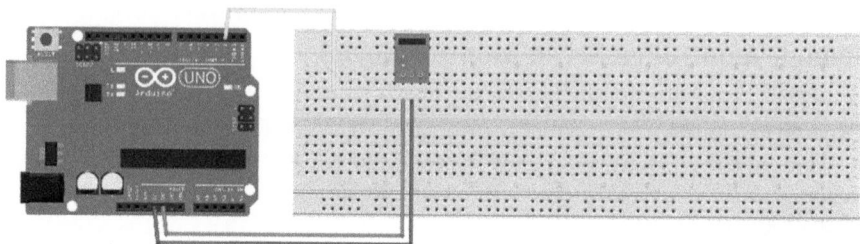

fritzing

5. After re-checking the wirings, connect the USB cable, and download the code.
6. Open the Serial Monitor (Tools > Serial Monitor). Each time you press a key on the remote, a code number should show up in the Serial Monitor. (For new IR remote, remove the plastic tab from the battery compartment). Note that -1 is separation between codes, it is not a code.
7. Decide which 2 keys you want to use for the next portion of this tutorial, then write down the codes that shows up in the Serial Monitor when you press those keys. Let's say the codes are X1 and X2. (e.g., X1 = 8295, X2 = -15811.)
8. After the first portion is complete and functionally tested, we will now replace the LED with a servo motor. If the plastic wing of the servo motor is not connected with the screw, connect it now so it is easy to see when the rotor moves.
9. Copy the code to a new Arduino sketch window.

```
#include <Servo.h>
Servo myservo; // create servo object
int pos = 0;   // servo position
void setup()
{
 myservo.attach(9); // attaches the servo on pin 9
}
void loop()
{
 for(pos = 0; pos < 180; pos += 1)  // forward
 {
  myservo.write(pos);
  delay(10);
 }
 for(pos = 180; pos>=1; pos-=1)   // backward
 {
  myservo.write(pos);
  delay(10);
 }
}
```

10. Now connect the servo motor on a different portion of the breadboard without taking out previous connections. The yellow wire of the motor is Signal (connects to Digital 9), the red wire is 5 V VDD, and the black (or brown) is the GND.

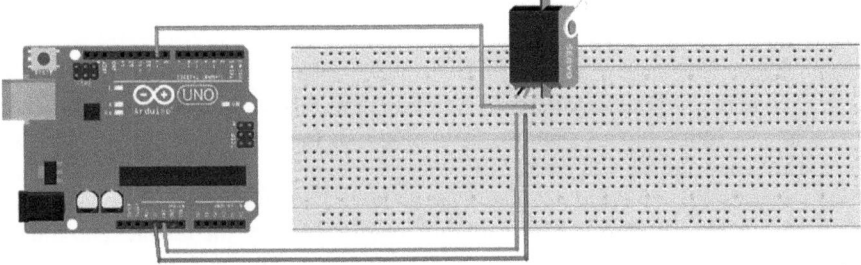

fritzing

11. After checking the wirings carefully, connect the USB cable.
12. Compile and download the code. If you execute this code, the motor should oscillate (forward and backward) from 0 to 180 degrees. If it is not, check hardware connectivity, or errors in copied code.
13. After the second portion is also complete, now we will combine the two codes. Type in the combined code as shown below:

```
#include <IRremote.h>
#include <Servo.h>
int RECV_PIN = 2;
IRrecv irrecv(RECV_PIN);
decode_results results;
int value;
Servo myservo; // create servo object
int pos = 0;   // servo position
void setup()
{
  irrecv.enableIRIn();
  myservo.attach(9); // attaches the servo on pin 9
}
void loop()
{
if (irrecv.decode(&results))// False if no data
{
value = results.value;// Store results
switch(value) {
  case X1:
  for(pos = 0; pos < 90; pos += 1)  // forward
  {
   myservo.write(pos);
   delay(50);
  }
  case X2:
```

```
for (pos = 90; pos>=1; pos-=1)     // backward
{
 myservo.write (pos);
  delay (50);
}
}
irrecv.resume (); // Ready to receive the next value
}
}
```

Note that you need to replace X1 and X2 in this code with your two codes that you found in first portion of this tutorial. (e.g., the code for X1 will be "8295" and for X2 will be "-15811.")

14. Connect USB cable and download the code. If everything is done properly, the servo motor should rotate forward with one keypress of the remote, and rotate backward with the other keypress of the remote. Note that the servo motor returns to rest state after completion of rotation. It might quickly go to starting point of the rotation if it is not starting from rest state.

Expected output/observation: The servo motor should turn one direction when the first assigned key is pressed, and turn the other way when the second assigned key is pressed.

Additional explorations:

(a) Can you find some new rotation schemes that always start from the rest state?
(b) Can you incorporate more keys to turn ON and OFF an LED along with this code?
(c) Can you update the code so that at the beginning of the running (initialization sequence), the servo rotates fully clockwise, then fully counterclockwise and the LED turns OFF? Can you incorporate another key in the remote so that when that key is pressed, the initialization sequence is run?

Appendix C: A Complete ES Project

I do not fear computers. I fear lack of them.—Isaac Asimov, Pioneer computer visionary

A Remote-Controlled Car

This project uses a DC motor toy car to be converted to remote control car with Arduino Uno and IR remote control. The required hardware setup is shown below.

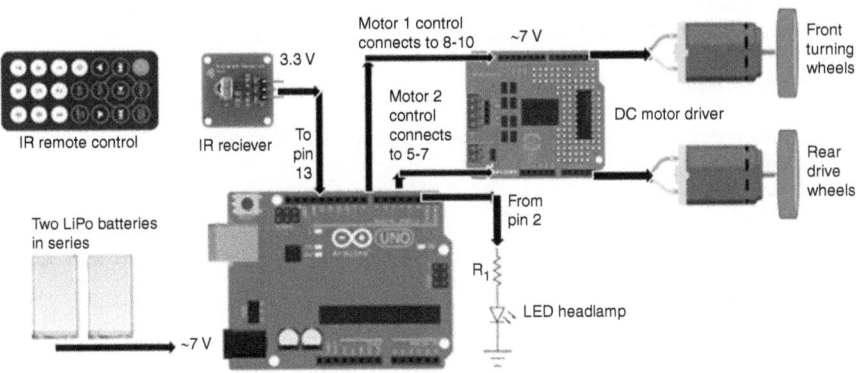

The Arduino Sketch code is given below.

```
#include "IRremote.h"
//include <Servo.h>
#include "ServoTimer2.h"
/*----( Declare Constants )----*/
int receiver = 13;
// pin 1 of IR receiver to Arduino digital pin 13
int IRbut = 99; // 99 for do nothing
int lastIRbut = 99;
```

© Springer Nature Switzerland AG 2021
B. I. Morshed, *Embedded Systems – A Hardware-Software Co-Design Approach*,
https://doi.org/10.1007/978-3-030-66808-2

```
int movespeed = 70; // change speed between 0 to 255
int turnspeed = 100; // change speed between 0 to 255
int left = 0; // servo motor left 0 degree
int right = 180; // 180 deg
int center = 90; // 90 deg
int pos = 0;    // variable to store the servo position
int ServoPin = 3; // Not used
// Servo control is connected in analog pin 0
//Servo myservo;
// create servo object to control a servo
ServoTimer2 myservo;
/*----( Declare objects )----*/
IRrecv irrecv(receiver);
// create instance of 'irrecv'
decode_results results;
// create instance of 'de-code_results'
/*----( Declare Variables )----*/
// connect motor controller pins to Arduino digital pins
// motor one
int enA = 10;
int in1 = 9;
int in2 = 8;
// motor two
int enB = 5;
int in3 = 7;
int in4 = 6;
/*----( Declare User-written Functions )----*/
void translateIR() // takes action based on IR code received
// describing Car MP3 IR codes
{
 switch(results.value)
 {
// codes verified for ELEGOO remote by BM on 11 Dec 2018
 case 0xFFA25D:
  IRbut = 10;
//  Serial.println(" OnOff      ");
  break;
 case 0xFFE21D:
  IRbut = 11;
//  Serial.println(" Funt      ");
  break;
 case 0xFFE01F:
  IRbut = 21;
//  Serial.println(" CH-       ");
  break;
 case 0xFF906F:
  IRbut = 22;
//  Serial.println(" CH+       ");
  break;
 case 0xFF22DD:
  IRbut = 31;
//  Serial.println(" PREV      ");
  break;
```

```
case 0xFFC23D:
 IRbut = 32;
// Serial.println(" NEXT       ");
 break;
case 0xFF02FD:
 IRbut = 33;
// Serial.println(" PLAY/PAUSE   ");
 break;
case 0xFFA857:
 IRbut = 41;
// Serial.println(" VOL-       ");
 break;
case 0xFF629D:
 IRbut = 42;
// Serial.println(" VOL+       ");
 break;
case 0xFF9867:
 IRbut = 43;
// Serial.println(" EQ        ");
 break;
case 0xFF6897:
 IRbut = 0;
// Serial.println(" 0        ");
 break;
case 0xFF30CF:
 IRbut = 1;
// Serial.println(" 1        ");
 break;
case 0xFF18E7:
 IRbut = 2;
// Serial.println(" 2        ");
 break;
case 0xFF7A85:
 IRbut = 3;
// Serial.println(" 3        ");
 break;
case 0xFF10EF:
 IRbut = 4;
// Serial.println(" 4        ");
 break;
case 0xFF38C7:
 IRbut = 5;
// Serial.println(" 5        ");
 break;
case 0xFF5AA5:
 IRbut = 6;
// Serial.println(" 6        ");
 break;
case 0xFF42BD:
 IRbut = 7;
// Serial.println(" 7        ");
 break;
```

```
  case 0xFF4AB5:
   IRbut = 8;
 //  Serial.println(" 8          ");
   break;
  case 0xFF52AD:
   IRbut = 9;
 //  Serial.println(" 9          ");
   break;
  default:
    IRbut = 99; // do nothing
 //  Serial.println(" other button  ");
   break;
 }
 // delay(500); // for serial monitor
 //delay(10); //otherwise
 } //END translateIR
 //////////////////////////// setup ////////////////////////////
 void setup() {
 // set all the motor control pins to outputs
 pinMode(enA, OUTPUT);
 pinMode(enB, OUTPUT);
 pinMode(in1, OUTPUT);
 pinMode(in2, OUTPUT);
 pinMode(in3, OUTPUT);
 pinMode(in4, OUTPUT);
 // for IR receiver remote
 // Serial.begin(9600);
 // Serial.println("IR Receiver Raw Data + Button Decode Test");
  irrecv.enableIRIn(); // Start the receiver
 // port setup
 DDRD |= 0b00000100;
 // set port D pin 2 as output, leave other pins untouched
 // for input, use & masked 0
 // myservo.attach(ServoPin);
 // attaches the servo on pin 12 to the servo object
 /* digitalWrite(ServoPin, LOW); // ensure servo off
  // initial movement of servo
  // move servo center
  myservo.write(center);
 // tell servo to go to position
  delay(100);
 // waits for the servo to reach the position
   PORTD ^= 0b00000100; // turn on
   // move servo left
  myservo.write(left);
 // tell servo to go to position
  delay(500);
 // waits for the servo to reach the position
  // move servo right
  myservo.write(right);
 // tell servo to go to position
  delay(500);
 // waits for the servo to reach the position
```

```
  // move servo center
  myservo.write(center);
  // tell servo to go to position
  delay(100);
  // waits for the servo to reach the position
    PORTD ^= 0b00000100; // turn off
    */
}
//////////////////////// End Setup ////////////////////////////
void slow1()
{
// this function will run the motors
// in both directions at a slow speed
// turn on motor A
digitalWrite(in1, HIGH);
digitalWrite(in2, LOW);
// set speed to 200 out of possible range 0~255
analogWrite(enA, 50); //right 40 lowest
// turn on motor B
digitalWrite(in3, HIGH);
digitalWrite(in4, LOW);
// set speed to 200 out of possible range 0~255
analogWrite(enB, 50); // left 50 lowest
delay(1000);
// now change motor directions
digitalWrite(in1, LOW);
digitalWrite(in2, HIGH);
digitalWrite(in3, LOW);
digitalWrite(in4, HIGH);
delay(1000);
// now turn off motors
digitalWrite(in1, LOW);
digitalWrite(in2, LOW);
digitalWrite(in3, LOW);
digitalWrite(in4, LOW);
}
///////////// Selected functions/modes //////////////////////////
void light() //10 onoff
{
  PORTD ^= 0b00000100; // toggle led
}
void dance() //11 funct
{
  light(); delay(500); light(); delay(200); //L2 <- Leven to end with
the same starting condition
  forward(); delay(500); stopcar();
  turnleft(); delay(5000); stopcar();
  light(); delay(500); light(); delay(200); //L2
  turnright(); delay(5000); stopcar();
  backward(); delay(500); stopcar();
  nudgeleft(); light(); delay(500); light(); delay(200); nudg-eright
(); //L2
  nudgeright(); light(); delay(500); light(); delay(200); nudge-left
(); //L2
  forward(); delay(400); stopcar();
```

```
    light(); delay(200); light(); delay(200); light(); delay(200);
light(); delay(200); //L4
    light(); delay(200); light(); delay(200); light(); delay(200);
light(); //L4
    backward(); delay(400); stopcar();
  }
  void turnleft() //21 ch-
  {
   // move servo left
    myservo.attach(ServoPin);
   // attaches the servo on pin 12 to the servo object
    myservo.write(left);
   // tell servo to go to position
    delay(100);
   // waits for the servo to reach the po-sition
    myservo.detach();
   // detaches the servo on pin 12 to the servo object
    // motor A forward
  digitalWrite(in1, LOW);
  digitalWrite(in2, HIGH);
   // motor B backward
  digitalWrite(in3, HIGH);
  digitalWrite(in4, LOW);
   // Move motors, set speed to 200 out of possible range 0~255
  analogWrite(enA, turnspeed); //right 40 lowest
  analogWrite(enB, turnspeed); // left 50 lowest
  }
  void turnright()  //22 ch+
  {
    // move servo right
   myservo.write(right);
   // tell servo to go to position
    delay(100);
   // waits for the servo to reach the po-sition
    // motor A backward
  digitalWrite(in1, HIGH);
  digitalWrite(in2, LOW);
   // motor B forward
  digitalWrite(in3, LOW);
  digitalWrite(in4, HIGH);
   // Move motors, set speed to 200 out of possible range 0~255
  analogWrite(enA, turnspeed); //right 40 lowest
  analogWrite(enB, turnspeed); // left 50 lowest
  }
  void nudgeleft() //31 prev
  {
   // motor A forward
  digitalWrite(in1, LOW);
  digitalWrite(in2, HIGH);
   // motor B backward
  digitalWrite(in3, HIGH);
  digitalWrite(in4, LOW);
   // Move motors, set speed to 200 out of possible range 0~255
```

```
analogWrite(enA, turnspeed); //right 40 lowest
analogWrite(enB, turnspeed); // left 50 lowest
delay(400); // turn a small amount
 // turn off motors
analogWrite(enA, 0); //right 40 lowest
analogWrite(enB, 0); // left 50 lowest
digitalWrite(in1, LOW);
digitalWrite(in2, LOW);
digitalWrite(in3, LOW);
digitalWrite(in4, LOW);
// continue if forward or backward
 switch (lastIRbut) {
  case 42: forward(); IRbut = 42; break;
  case 41: backward(); IRbut = 41; break;
  default: break;
 }
}
void nudgeright() //32 next
{
  // motor A backward
digitalWrite(in1, HIGH);
digitalWrite(in2, LOW);
// motor B forward
digitalWrite(in3, LOW);
digitalWrite(in4, HIGH);
// Move motors, set speed to 200 out of possible range 0~255
analogWrite(enA, turnspeed); //right 40 lowest
analogWrite(enB, turnspeed); // left 50 lowest
delay(400); // turn a small amount
 // turn off motors
analogWrite(enA, 0); //right 40 lowest
analogWrite(enB, 0); // left 50 lowest
digitalWrite(in1, LOW);
digitalWrite(in2, LOW);
digitalWrite(in3, LOW);
digitalWrite(in4, LOW);
// continue if forward or backward
  switch (lastIRbut) {
  case 42: {forward(); IRbut = 42;} break;
  case 41: {backward(); IRbut = 41;} break;
  default: break;
 }
}
void stopcar() //33 play/pause
{
 // to stop immediately, put reverse force
 switch (lastIRbut) {
  case 42: backward(); delay(20); break;
  case 41: forward(); delay(20); break;
  case 21: turnright(); delay(10); break;
  case 22: turnleft(); delay(10); break;
  default: break;
 }
```

```
 // turn off motors
analogWrite(enA, 0); //right 40 lowest
analogWrite(enB, 0); // left 50 lowest
digitalWrite(in1, LOW);
digitalWrite(in2, LOW);
digitalWrite(in3, LOW);
digitalWrite(in4, LOW);
 // move servo center
 myservo.write(center);
// tell servo to go to position
 delay(100);
// waits for the servo to reach the position
}
void backward() //41 vol-
{
// motor A backward
digitalWrite(in1, HIGH);
digitalWrite(in2, LOW);
// motor B backward
digitalWrite(in3, HIGH);
digitalWrite(in4, LOW);
// Move motors, set speed to 200 out of possible range 0~255
analogWrite(enA, movespeed); //right 40 lowest
analogWrite(enB, movespeed); // left 50 lowest
}
void forward() //42 vol+
{
 // move servo center
 myservo.write(center);
// tell servo to go to position
 delay(100);
// waits for the servo to reach the position
// motor A forward
digitalWrite(in1, LOW);
digitalWrite(in2, HIGH);
// motor B forward
digitalWrite(in3, LOW);
digitalWrite(in4, HIGH);
// Move motors, set speed to 200 out of possible range 0~255
analogWrite(enA, movespeed); //right 40 lowest
analogWrite(enB, movespeed); // left 50 lowest
}
void backbrief() //43 EQ
{
 backward();
 delay(700); // set how long is "brief"
 // turn off motors
analogWrite(enA, 0); //right 40 lowest
analogWrite(enB, 0); // left 50 lowest
digitalWrite(in1, LOW);
digitalWrite(in2, LOW);
digitalWrite(in3, LOW);
digitalWrite(in4, LOW);
```

```
}
//////////////// Main loop starts here ///////////////////
void loop()
{
  if (irrecv.decode(&results))
// have we received an IR signal?
  {
//   Serial.println(results.value, HEX);
//UN Comment to see raw values
    translateIR();
//Serial.println(IRbut); // to check the code
  switch(IRbut) {
    case 10: light(); break; // onoff
    case 11: dance(); break; // funct
    case 21: turnleft(); break; //ch-
    case 22: turnright(); break; //ch+
    case 31: nudgeleft(); break; //prev
    case 32: nudgeright(); break; //next
    case 33: stopcar(); break; // play/pause
    case 41: backward(); break; // vol-
    case 42: forward(); break; // vol+
    case 43: backbrief(); break;// EQ, short backup
    default: break; // unknown or 99
  }
  lastIRbut = IRbut; // save last IR code
  //check for code after at least 100 ms
  delay(100);
// digitalWrite(ServoPin, LOW); // ensure servo off
//Servo::refresh();
  irrecv.resume(); // receive the next value
}
//else
// IRbut = 99; // ensure action is taken only once
//slow1(); // motor motion
//delay(1000);
}
//////////////////////// End of code ////////////////////////
```

Appendix D: Sample ES Assignments

The more you know, the more you realize you know nothing.— Socrates

Embedded Systems Market and UML

Problem 1

Assume that you are leading a team that is assigned to design a remote-controlled robotic humanoid system. The system composes of two distinct units: a robot unit that can move freely on a set of wheels, and a remote control unit. The two units communicate via a wireless link. The minimum requirements for the robot unit are:

- Basic movements: Move Forward, Move Backward, Turn Left, Turn Right.
- Human-machine interfacing (HCI): (i) Greet when turned on and when any person approaches the robot, (ii) Transmit bidirectional audio and unidirectional video from the robot to the remote control unit for real-time monitoring and control, (iii) an audio-video recorder in the robot unit for data logging.

Provide the followings in terms of UML:

(a) A list of **Structural Things** that is required at the minimum for the system.
(b) List at least **4 use cases** for the system.
(c) Draw the following UML diagrams:

1. Class diagram
2. Deployment diagram
3. Use-case Diagram (corresponding to 1b)
4. StateChart diagram (corresponding to b)
5. Sequence diagram (corresponding to 1b)

© Springer Nature Switzerland AG 2021
B. I. Morshed, *Embedded Systems – A Hardware-Software Co-Design Approach*,
https://doi.org/10.1007/978-3-030-66808-2

Problem 2

(a) For an embedded system product with a market life of 5 years, compare the revenues for an on-time product and a delayed product by (i) 6 months and (ii) 10 months. Assume the triangular approximation model for the market window and the revenue peaks at $100K.
(b) Compare the percentage loss of revenue for the two delayed products.

Problem 3

For an embedded system product, the NRE cost and unit cost are the following for the four technologies:

Technology	NRE expense	Unit cost
Semi-custom VLSI	$200,000	$5
ASIC	$50,000	$10
Programmable FPGA	$15,000	$20
Microcontroller	$10,000	$15

(a) Calculate total per-product cost for production volumes of 100, 1k, 10k, and 100k units.
(b) Plot these data from (a) in a single graph with log scale for per-product cost and draw piecewise linear lines for each technology. Then, determine the best choice of technologies for these production volumes (100, 1k, 10k, and 100k units) to achieve the lowest per-product cost. Also plot total cost for these product volumes in a separate log-log graph.
(c) From the per-product cost plot in the (b), estimate the range of production volumes for which each of these technologies is financially optimal.
(d) List 3 other considerations in addition to per-product cost that might affect the choice of technology.

Traffic Light Controller with Arduino

In this assignment, you will develop a simple traffic light control system for a four-way street intersection using Arduino Uno board. At the minimum, you must prepare two sets of traffic lights (each set consisting of a Red, Yellow (Amber), and Green LEDs) of adjacent streets (or cross-streets, e.g., North and East) and two switches (push-type) for pedestrians, one for each of the streets (e.g., one for North and other for East). An example of circuit connection diagram for one set is shown below (left). You will need to prepare two sets of this circuit for this assignment. Digital output pins and digital input pins needs to be connected to your Arduino Digital pins (between pin 2 to 13).

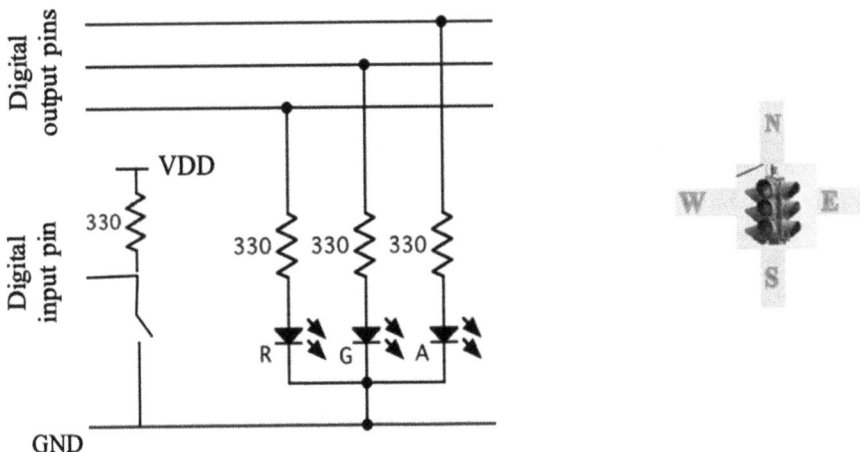

As a requirement for this assignment, you will need to prepare the circuit on a breadboard with proper connection to your Arduino and write a software code using Arduino Sketch to control the LEDs in proper sequence, and allow the switch press to modify the timing of the sequence. Normally (if none of the switches are pressed), the LEDs will turn on and off in this sequence: Red -> Green -> Amber -> Red, etc. with the given time delay mentioned later. As soon as one of the push buttons is pressed (on either of the streets), the corresponding Amber (yellow) LED will light up, followed by Red for that side of the street. The adjacent street must also have the opposite sequence (i.e., opposite to Red light is Green followed by Amber). After Red light time expires, then the sequence will continue according to normal timings. If the switch is pressed when the Red or Amber LED is ON, then the switch press will be ignored.

To construct this setup, you can follow this procedure:

(1) First make the LED circuits shown above. Use the breadboard to setup at least 2 sets of 3 LEDs (i.e., 6 LEDs in total: 2 Red, 2 Yellow, and 2 Green). Each set will consist of one red (R), one yellow/amber (A), and one green (G) LED. Each LED must have a series Resistance of 330Ω (or 220Ω) as shown.

(2) Connect the LEDs to different Digital I/O pins of the Arduino Uno (you can use any Digital pin between 2 to 13, e.g., for one set you can use pins 2, 3, & 4, and for the other set you can use pins 6, 7, 8). Refer to Tutorial 1 for further details of LED connections. (Note: *Do not* use digital pin 0 or 1.)

(3) Following the Tutorial 1 code, write a test code to turn ON/OFF (blink) each LED independently (this process ensures that your LEDs are connected

properly). Note that you should change the digital pin number in your test code to verify proper connection of each of the LEDs one by one (do not connect multiple LEDs on a single digital pin). You can use this type of (partial) checking for correctness of your (partial) hardware setup in any project circuit setup testing to avoid later pitfall (a strategy known as **Divide and Conquer!**).

(4) Next, you can start programming the state machine. Here you can use simple delay() statements. For the timing of the LEDs (R->G->A->R->G->A, etc.), use this following sequence and times for delays.

Note: The delay of red light needs to be equal to the total delay of green and amber lights – this is a constraint of the timing for traffic signal lights.

You also need to ensure the timing of the adjacent-street is operating opposite way (see the timing diagram below), to allow proper traffic flow on both streets. An example timing diagram for the two sets of lights are shown below. For instance, Traffic light set 1 can be for North (N) direction, and Traffic light set 2 can be for East (E) direction.

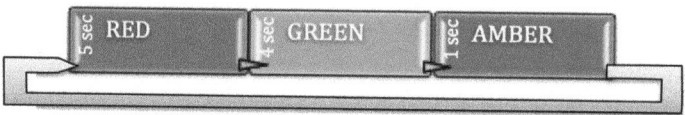

(5) Implement this first with properly thought-out delay statements in Arduino Sketch. After completing the code, upload the code to your Arduino and test LEDs for functionality.

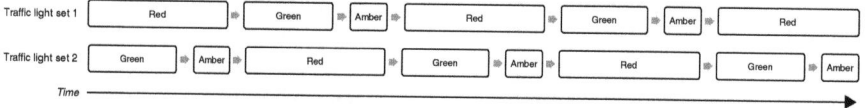

(6) Now, you can add two pedestrian switches. A pedestrian can press this switch to immediately change the light to red so that he/she can cross the street. Set up the switches as given in the schematic above to free Digital pins (between 2 and 13, e.g., digital pin 5 for one side and digital pin 9 for other side). Make sure these Digital pins are set as INPUT pin in the setup(), for example *pinMode (5, INPUT)*. Also, you need to read these inputs as digital logic value in loop(), for example *val1 = digitalRead(5)*. You can then use *val1* as Boolean logic (TRUE/FALSE) in your *if* statements. Note that the circuit shown above will produce logic 1 (+5 V) when the switch is not pressed and will produce logic 0 (0 V) when pressed (a setup known as Active Low).

(7) Now you will need to update your code to accommodate this switch presses (i.e., if the LED color is changing with proper timing as shown above when no switch is pressed, but if a switch is pressed with corresponding LED in Green, the corresponding street LED needs to change to Amber followed by Red, while the adjacent street LED needs to change to Green when other street light becomes

Red. The sequence will follow thereafter normally according to the timing diagram above.)

In the assignment report, provide the followings:

(a) Briefly summarize the work including your objective, procedure, and key results.
(b) Provide your code for this traffic light controller with sufficient comments to ensure the code is understandable.
(c) Draw a StateChart for your code.
(d) Explain how you implemented the pedestrian switch to respond when pressed? Was it instantaneous? What is the best-case delay and worst-case delay from the time when the switch was pressed and the time when the Amber (yellow) LED turned on? Can this be improved? How?
(e) Provide snapshots confirming successful download of the code from computer to Arduino board.
(f) Provide a few pictures as evidence of successful prototyping of the hardware and operation of the code with various combination of LEDs ON/OFF.
(g) Did you encounter any technical challenge? How did you resolve it?

Gesture-Controlled Servo Motor with Arduino

In this assignment, you will develop a gesture-controlled servo motor using Arduino Uno kit. To detect gesture, you can use hand movement toward a light sensor as input, and to produce control signals for servo motor, we will use LDR values (after scaling) for decrease of light on the sensor to turn the motor counterclockwise ($0°$), and increase of light on the sensor to turn to motor clockwise ($180°$). Instead of your hand, you can also use your palm, an opaque piece of plastic or paper, or an opaque object that provides sufficient shade on LDR. Some procedural steps to develop the light sensor portion and the servo motor portion are provided below as a guideline. Also, you can consult Tutorial 2 (B.2) for LDR operation and Tutorial 4 (B.4) for servo motor operation (Appendix B).

Test Procedure for Light Sensor

Component required:
An LDR (light dependent resistor), a POT (potentiometer or variable resistor), a resister (1 kΩ), a breadboard, and some jumper wires.

Steps to test LDR:

(1) First, collect the light dependant resistor (LDR) (Fig. 1 left), a pot (potentiometer) (Fig. 1 right), and a 1 kΩ resistor from your Arduino kit. An LDR is a *photoresistor*, sometimes called a *photocell*. It is in a form of variable resistor whose resistance decreases (if negative coefficient, or vice versa) with exposure to light. In effect, an LDR is a *light sensor*.

(2) Now connect these in series as shown in Fig. 2 on a breadboard. Note, we are using Pin 1 and 2 of the pot (leaving Pin 3 open). Connect a jumper wire from Analog pin A0 to the LDR as shown.

(3) In this circuit, we have connected the pot to allow us to control sensitivity. By turning the pot, you can make the LDR more or less sensitive to light. The series resister of 1 kΩ is added to provide protection from over-current flow through the LDR when the resistance value of the pot is zero.

(4) Now, follow the first portion of tutorial 2 (B.2) codes and procedure to test out the LDR connection and operation. Using the same procedure, check if you receive the LDR data at the serial monitor display of the computer. Use the serial plotter, then adjust the pot such that when you move your hand close to the light sensor, you have a large deviation in the value. Record these values and do not change pot setting any further, after you are satisfied with sensitivity of the LDR. This will give you information regarding the range of digital values that you can scale to motor rotation.

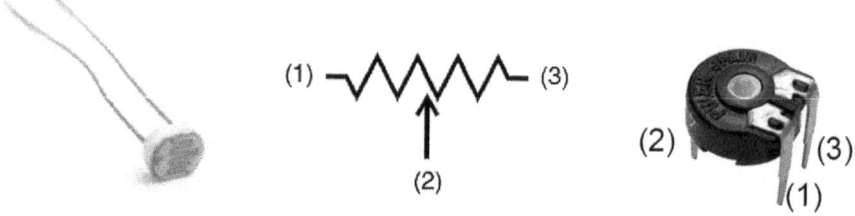

Fig. 1. (left) Photograph of an LDR. (right) A potentiometer (pot) schematic and an image showing pin numbers. Plug in the plastic handle to the pot if not already done.

Fig. 2. Circuit to connect the LDR output A0 to the Arduino Analog pin A0.

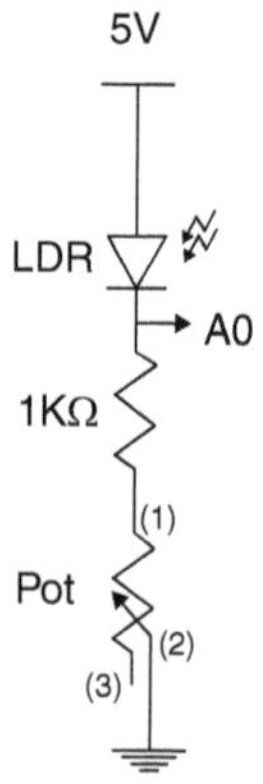

Test Procedure for Servo Motor

Components required:

A servo motor, 3 jumper wires.

Steps:

(1) Do not alter the hardware setup that you have done above. Add this new hardware setup as mentioned below without any change of the previous hardware (which will not affect the test described below).

(2) In the Arduino kit, find the servo-motor and a plastic rotor (it might be already connected) as shown below. If not connected, connect the rotor to the motor spindle using the supplied screw.

(3) Following the procedure from second portion of tutorial 4 (B.4), connect the servo motor to your Arduino board. Download the test code and ensure the servo motor is moving clockwise and counterclockwise.

(4) If you execute this code, the motor should oscillate from $0°$ to $180°$. If it is not, check hardware connectivity, or errors in copy of the code, compilation error, and proper execution of the code.

Note: If the servo motor is still not operating, check the wiring to ensure you have connected the correct terminals to the +5 V, Gnd, and Pin 9 of the Arduino board. Consult B.4 tutorial. The motor should follow these changes accordingly.

Task

With the hardware setup as above (both tests), write a new code such that the servo motor direction can be controlled by moving your hand closer or further from the light sensor. When the hand is very close to the LDR, the motor should be at an angle of $0°$, while when the hand is away, the motor should be at an angle of $180°$.

Note 1: You might need to fine-tune the delay of pulses provided to the servo motor and the scaling for digital data to motor rotation to properly control the motion of the servo motor for smooth transitions (i.e., as your hand approaches the sensor, the motor moves accordingly and vice versa).

Note 2: You might need to change the data format from LDR to control the servo motor angle.

Note 3: You might need to scale and curve fit the data from LDR to fit the range of servo motor angle ($0°$ to $180°$) as smooth transitions.

In the assignment report, provide the followings:

(a) Briefly summarize the work including your objective, procedure, and key results.

(b) Provide your code for this gesture-controlled servo motor with sufficient comments to ensure the code is understandable.

(c) Draw a sequence diagram for your code.

(d) Explain how you implemented the servo motor movement to correlate your hand gesture? Can this be improved to detect other types of motion of hand, such as side to side? How?

(e) Provide snapshots confirming successful download of the code from computer to Arduino board.
(f) Provide a few pictures as evidence of successful prototyping of the hardware and operation of the code with various conditions (e.g., hand near the LDR, hand away from the LDR).
(g) Did you encounter any technical challenge? How did you resolve it?

Closed Loop Feedback PID Controller with Arduino

In this assignment, you will develop a feedback controller for closed loop cyber physical system (CPS) that encompasses both cyber and physical domains. As a control algorithm, we will use PID (Proportional Integral Differential) algorithm. We will use a light sensor as input and an LED for output. These will interface the cyber and physical domains. PID is a simple controller system. If you are not familiar with PID controller, here is a short tutorial: https://en.wikipedia.org/wiki/PID_controller

Hardware Setup and Software Test

Component required:
An LDR (light dependent resistor), a POT (potentiometer or variable resistor), a red LED, a 220 (or 330) Ω resistor, a 10 kΩ resistor, a breadboard, and some jumper wires.
Steps:

(1) Connect the Arduino with the components as follow:

- Photoresistor in series with 10 kΩ resistor, then connect the open end of the photoresistor to 3.3 V and the open end of the resistor to 0 V (GND). Then connect the common point to A0 input.
- Connect one side of the POT to 3.3 V and the other side to 0 V (GND). Then connect the middle point to A1 input.
- Connect an LED in series with a resistor (220 Ω or 330 Ω) such that negative terminal of the LED connect to the resistor. Connect the open positive terminal of the LED to digital pin 9. Then connect the open side of the resistor to 0V (GND).

(2) An example of connection diagram on the breadboard is given below.

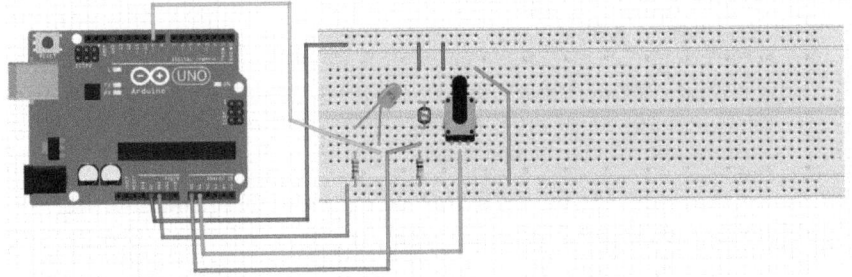

(3) The hardware setup might look like this. (Implemented differently on the breadboard compared to above diagram. Either is fine as long as the connection is proper.). Note that the LDR and the LED must face each other with a small gap as shown.

(4) Download the PID controller zip file from this link: https://github.com/br3ttb/ Arduino-PID-Library. Unzip the content on your computer Desktop (or in another suitable folder of your choice).

(5) From the Arduino software menu, click **Toolbar -> Sketch -> Import Library -> Add Library .zip**. Browse to the desktop (or the folder you have downloaded the zip file) and select **PID_v1**.

(6) With a blank code in Arduino software, click **Toolbar -> Sketch -> Import Library -> PID_v1**. This should insert the following text on the code.
    #include <PID_v1.h>

(7) Copy the rest of the code from the Provided Code given at the end of this document.

(8) Compile the code and upload it on the Arduino. You have to resolve any error prior to download, if any.

(9) If the code is correct and downloaded properly, LED should change brightness when you turn the POT.

Hint: If nothing is happening:

- *Try pressing the reset button of the Arduino board.*
- *Check the code in Arduino sketch and compare with the provided code to see if any line is missing, broken down to multiple lines, combined to previous line, or commented out.*

(10) Now, Ensure the **LED and LDR are heading each other (front to front)** very closely, and all wires are connected properly.

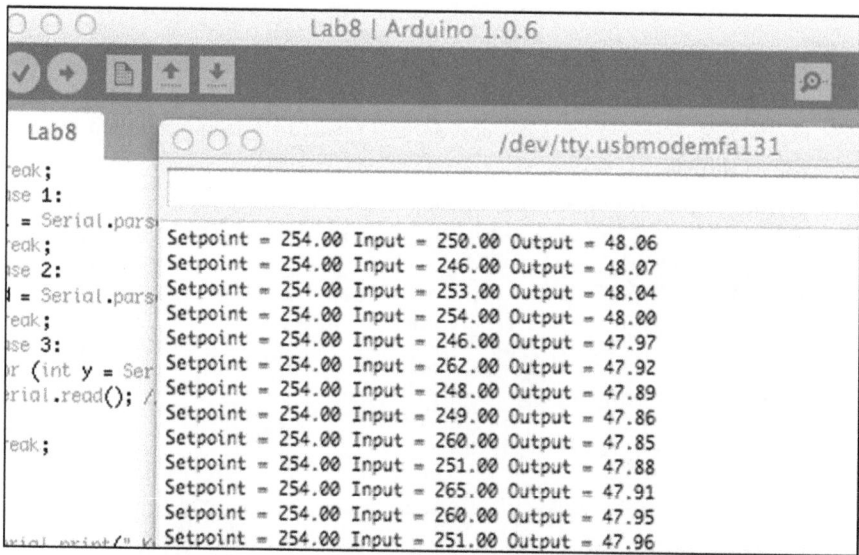

(11) Start the Serial monitor in Arduino Sketch. You should see data as shown below.

(12) In the Arduino Sketch software code, set the initial values for gains to **0 for Kp and Kd**, and **10 for Ki**. This will make the LED to have a constant voltage, and the Ki will turn up till it reaches 255 or the LED matches the desired set point.

(13) With the LED very close to your photo resistor, change the potentiometer setting. This should increase and decrease the brightness of the LED. If you move the LED away, the PID feedback loop should try to compensate and turn up the brightness to match the set point. Try inserting a paper in between the LED and LDR and observe. The LED should turn brighter.

(14) In the serial monitor, send in a command of "**0,0.5,0**" without the quotes. The Arduino should parse it, and set Ki to 0.5 instead of 10. Change the Pot setting to a low brightness, and repeat this step again, and observe any difference.

(15) DIRECT tells the PID which direction it is working in (refer to PID function in the Given Code). If, for example, the PID loop tries to make the LED dimmer when it should actually get brighter, we would change DIRECT to REVERSE.

Change the code to REVERSE, then upload again, and try the above steps to notice any change.

Notes on the code:

- *PID myPID(&Input, &Output, &Setpoint, Kp, Ki, Kd, DIRECT);*
  This line set up the PID library to use a PID process called myPID.
- Any time we call

```
myPID.Compute();
```

subsequently, myPID will take the variables Input, Setpoint, Kp, Ki, and Kd, and use them in its calculation, and then write whatever it determines to Output.

Task 1

Find answers to the following questions:

(i). *Explain your observation in Step 12. Why LED turns brighter when a paper is inserted between the LED and the LDR?*

(ii). *In step 13, do the following and note the observations:*
Keeping Kp and Kd to 0 and 0, respectively, change ki to 0, 0.5, 1, 5, 10
Keeping Kp and Ki to 0 and 10, respectively, change kd to 0, 0.5, 1, 5, 10
Keeping Kd and Ki to 0 and 10, respectively, change kp to 0, 0.5, 1, 5, 10

(iii). How does the rate of the PID loop reacts to (a) changes of environmental light and (b) changes to the set point? Draw a flow chart to explain this.

(iv). *How DIRECT (direction) effect the operation? What is the value for this? Change this DIRECT to REVERSE, and observe what happens. Explain your observation.*

Task 2

In the assignment report, provide the followings:

(a) Briefly summarize the work including your objective, procedure, and key results.
(b) Describe the control system that you have implemented, and explain each term of PID.
(c) Provide the complete software code with sufficient comments.
(d) Provide the answers from the questions listed under Task 1 (i) to (iv).
(e) Provide snapshots confirming successful download of the code.
(f) Provide a few pictures showing your setup and proper functionality.
(g) Did you encounter any technical challenge? How did you resolve it?

Provided Code:
Code for PID Controller

```
//This code is from a lab of Portland State University
//From: Bret Comnes & A. La Rosa
//The PID library is provided by Beauregard,
//PIDLibrary, 2013
```

```
//Creative commons usage
#include <PID_v1.h>
const int photores = A0; // Photo resistor input
const int pot = A1; // Potentiometer input
const int led = 9; // LED output
double lightLevel;
// variable that stores the incoming light level
// Tuning parameters
float Kp=0; //Initial Proportional Gain
float Ki=10; //Initial Integral Gain
float Kd=0; //Initial Differential Gain
double Setpoint, Input, Output;
//These are just variables for stor-ingvalues
PID myPID(&Input, &Output, &Setpoint, Kp, Ki, Kd, DIRECT);
// This sets up our PDID Loop
//Input is our PV
//Output is our u(t)
//Setpoint is our SP
const int sampleRate = 1;
// Variable that determines how fast our PID loop runs
// Communication setup
const long serialPing = 500;
//This determines how often we ping our loop
// Serial pingback interval in milliseconds
unsigned long now = 0;
//This variable is used to keep track of time
// placehodler for current timestamp
unsigned long lastMessage = 0;
//This keeps track of when our loop last
//spoke to serial last message timestamp.
void setup() {
 lightLevel = analogRead(photores); //Read in light level
 Input = map(lightLevel, 0, 1024, 0, 255);
 //Change read scale to analog
 //out scale
 Setpoint = map(analogRead(pot), 0, 1024, 0, 255);
 //get our setpoint from our pot
 Serial.begin(9600); //Start a serial session
 myPID.SetMode(AUTOMATIC); //Turn on the PID loop
 myPID.SetSampleTime(sampleRate); //Sets the sample rate
 Serial.println("Begin"); // Hello World!
 lastMessage = millis(); // timestamp
}
void loop() {
 Setpoint = map(analogRead(pot), 0, 1024, 0, 255);
 //Read our setpoint
 lightLevel = analogRead(photores); //Get the light level
 Input = map(lightLevel, 0, 900, 0, 255);
 //Map it to the right scale
 myPID.Compute(); //Run the PID loop
 analogWrite(led, Output);
 //Write out the output from the PID loop
 //to our LED pin
```

```
now = millis(); //Keep track of time
if(now - lastMessage > serialPing) {
//If it has been long enough give us
//some info on serial
// this should execute less frequently
// send a message back to the mother ship
  Serial.print("Setpoint = ");
  Serial.print(Setpoint);
  Serial.print(" Input = ");
  Serial.print(Input);
  Serial.print(" Output = ");
  Serial.print(Output);
  Serial.print("\n");
 if (Serial.available() > 0) {
//If we sent the program a command deal
//with it
 for (int x = 0; x < 4; x++) {
  switch (x) {
    case 0:
      Kp = Serial.parseFloat();
      break;
    case 1:
      Ki = Serial.parseFloat();
      break;
    case 2:
      Kd = Serial.parseFloat();
      break;
    case 3:
      for (int y = Serial.available(); y == 0; y--) {
        Serial.read(); //Clear out any residual junk
      }
      break;
  }
 }
 Serial.print(" Kp,Ki,Kd = ");
 Serial.print(Kp);
 Serial.print(",");
 Serial.print(Ki);
 Serial.print(",");
 Serial.println(Kd); //Let us know what we just received
 myPID.SetTunings(Kp, Ki, Kd);
//Set the PID gain constants
//and start running
 }
 lastMessage = now; //update the time stamp.
 }
}
```

Analog Filter Design with Analog Filter Wizard and KiCad

In this assignment, you will design an analog filter using Analog Filter Wizard (by Analog Devices). https://tools.analog.com/en/filterwizard/

Problem 1

Using Analog Filter Wizard tool, design an active band pass filter using an op-amp that meets these specifications:

Center frequency (f_c) = 1 kHz
Pass-band gain (A) = 10 dB
Pass-band BW (-3 dB) = 100 Hz
Stop band (-40 dB) = 20 kHz
Number of stages = 1 (Fewest stages)
Filter Type = Chebyshev
Filter order = 2nd
Power supply: +Vs = 5 V, -Vs = 0 V
Optimization: Low noise
Use Standard component tolerances (as shown by default).

Task 1
Use Analog Devices Filter Wizard (ADFW) to design a filter circuit. Determine the frequency response by looking at Magnitude (dB) response and check if it matches the given specifications. Save your circuit design and results.

Problem 2

Use KiCad to design the schematic that you have done in Task 1, and perform simulations (transient and AC). Make sure to use pspice opamp for both units for this simulation and the "opamp model for LM258.sub" SPICE model. Change the sub-circuit declaration as follows to match the pin-sequence of the sub-circuit with that of KiCad opamp model:

```
.SUBCKT LM258_ON  24 2 1  12 11
```

Task 2
Using KiCad, complete the design of the circuit design with schematic and simulate the design to obtain results that matches with Task 1.

Assignment report:
In the assignment report, provide the followings:

(1) **From Task 1 - Analog Filter Wizard (AFW):**
   Snapshots of:

   (a) Circuit diagram,
   (b) Magnitude plot,
   (c) Phase plot,
   (d) Step response,
   (e) Power (of each stage),
   (f) Noise spectral density.

(2) **From Task 2 - KiCad:**

   (a) Snapshots of the complete Schematic diagram with sufficient zooms so that component values are legible.
   (b) Transient signal plots for 10 Hz, 100 Hz, 1 KHz, and 10 KHz with input signal of 10 mV. Choose simulation time such that each plot shows 10 full cycles of the input signal and the output signal on each plot. Include these plots in your report with notes of simulation frequency, input voltage range and output voltage range.
   (c) Transient signal plots for 1 mV, 10 mV, 100 mV, and 1 V with input signal frequency of 1 kHz. Each plot must show 10 full cycles of both the input signal and the output signal on the same plot. Include these plots in your report with notes of simulation frequency, input voltage range and output voltage range.
   (d) Use AC simulation to find out frequency response (Vout) from 1 Hz to 20 kHz. Compare this plot with the AFW magnitude plot by providing the plots side by side. Explain any difference you observe.

PCB Design of an Analog Filter Circuit using KiCad

In this assignment, you will need to design a PCB for the analog filter circuit that you have created in KiCad for the last assignment.

Task

Design a **2-layer** through-hole type (THT) PCB design using KiCad software (i.e., PCB layout editor) for the filter schematic that you have developed in the last assignment. For help in the design steps, see "KiCad Getting Started" document (pages 18, 21-28), in addition to Chapter 3 - Hardware and Appendix A.

Use thorough hole footprints (i.e., THT type) for all components and connectors. Use the following guidelines for the footprint assignment of your circuit components:

(1) Use R-PDIP-T8 footprint for the op-amps. You can use the standard 8 PIN DIP type footprint (e.g., **Package_DIP:DIP-8_W10.16mm**).

(2) For through-hole type registers (that you use in Breadboard circuit), use footprint of 1/8W (e.g., **Resistor_THT: R_axial_DIN0204_L3.6mm_D1.6mm_P7.62mm_Horizontal**),

(3) For capacitors, select Radial footprint (e.g., **Capacitor_THT: CP_Radial_D5.0mm_P2.00mm**) if the capacitor value is larger (C ≥ 1 μF) for Electrolyte type capacitors, while use Disk capacitor footprint (e.g., **Capacitor_THT:C_Disc_D3.0mm_W2.0mm_P2.50mm**) for smaller capacitors (C < 1 μF) for Ceramic type capacitors.

(4) For both voltage sources (V1 and V2), and the load resistance (R connected to Vout), using 2-terminal Pin Headers. For pin header connectors, use 2.54 mm (i.e., 0.1 inch) pitch (spacing) vertical headers (e.g., **Connector_PinHeader_2.54mm:PinHeader_1x02_P2.54mm_Vertical**).

Using PCB layout editor tool, import your circuit (rats-nest), space them out in your design area (placement), and route the traces to complete designing the PCB (routed design). Note that if there is any white links left, your routing is not complete. Also note the trace of two top layer cannot intersect (i.e., short circuit), which must be avoided by using Via (shortcut key "V") to route the trace to bottom layer than back to top layer using another Via before reaching the destination terminal.

For report snapshots, zoom in such a way that the design occupies most space in the computer screen (i.e., Zoom Full).

Assignment report:

In the report document, provide the followings with clear descriptions (along with snapshots as needed):

(a) Provide screenshot of the circuit (schematic) design that you implemented.
(b) Provide screenshot of the initial PCB design (i.e., rats-nest form).
(c) Provide screenshot after placement (before routing).
(d) Provide screenshot of final PCB design after routing is complete. Note that there cannot be any white links of rats-nest left in the design.
(e) Provide screenshot of Design Rule Check (DRC) test result showing that there are no errors.
(f) Provide screenshots of 3D view for both top side and bottom side.

Timer and Interrupt Experiment with Arduino

In this assignment, we will investigate using timer and interrupt of Arduino Uno board (with MCU: ATmega 328). This microcontroller has 3 timer units (Timer/Counter 0, Timer/Counter 1, Timer/Counter 2). For details of these timers, see ATmega328 datasheet, sections 12-15. In this exercise, we will use PCINT interrupt. For details of this interrupt, see ATmega328 datasheet, sections 9-10. We will also explore sleep modes. For sleep mode information, refer to ATmega328 datasheet, section 7.

Hardware Setup:

Component required:

A red LED, a 220 (or 330) Ω resistor, a push switch, a 1 kΩ resistor, a breadboard, and jumper wires.

Initial Test Procedure:

Setup an Arduino Uno board hardware in the following way with a push switch from 3.3V to Pin 10 with a 1 kΩ series resistor to ground, and an LED in series with a 220 Ω (or 330 Ω) resistor connected to Pin 13.

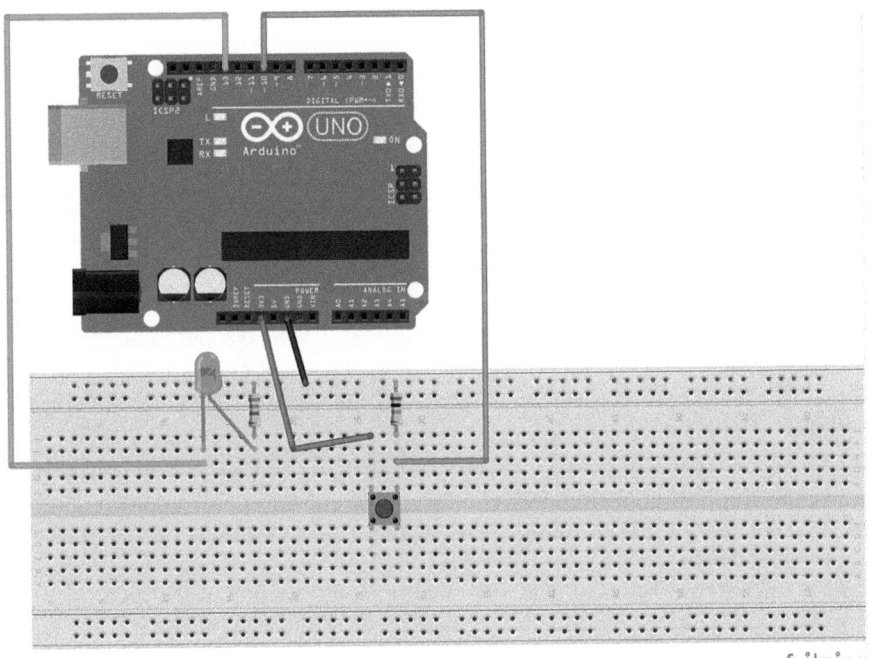

Copy the code from Code A (below) to Arduino Sketch. Compile and download it on Arduino. If your setup is correct, the LED should turn ON when the push switch is pressed, otherwise be OFF.

Task 1

For this assignment you must use both PCINT interrupt and Timer2 interrupt. An incomplete code is provided in Code B (below) to assist you in programing. *The example is by no means complete or provides all the things you need to consider, but just an elaborated hint.*

Your task is to program Arduino interrupt PCINT0 to Pin 10 so that when the push button is pressed, the Timer/Counter 2 will be enabled. It must produce a delay of 2 seconds. After 2 seconds, the LED must turn ON. Then Timer/Counter 2 will count another 3 seconds. After that the LED will turn OFF. For timer calculation, use Arduino default clock speed of 16 MHz.

Start the testing with highest power sleep mode (IDLE). When functional, try to find the possible *lowest power sleep mode*.

If you get an error message for missing avr/sleep.h, download "sleep.h" from AVR library (https://github.com/vancegroup-mirrors/avr-libc/tree/master/avr-libc/include/avr) and copy to Arduino Library folder on your computer.

Task 2

In the assignment report, provide the followings:

(a) Briefly summarize the work including your objective, procedure, and key results.
(b) Describe how you determined the contents (values) of each register in the given Code B.
(c) Explain why Timer 2 is used. What are the key advantage and disadvantage of using Timer 1?
(d) Which lowest power sleep mode you were able to use? Why not other sleep modes? Explain.
(e) Provide the final completed code (of given Code B) with sufficient comments to ensure the code is understandable.
(f) Provide snapshots confirming successful download of the code from computer to Arduino board.
(g) Provide a few pictures as evidence of successful operation of the code.
(h) Did you encounter any technical challenge? How did you resolve it?

Code A:
Code to test hardware setup:

```
void setup() {
  pinMode(10,INPUT);
  pinMode(13,OUTPUT);
}
void loop() {
  digitalWrite(13,LOW);
  if (digitalRead(10)) {
    digitalWrite(13,HIGH);
  }
}
```

Code B:
Partially completed C code is given below. Note that the Serial Monitor is only provided for debugging purpose.

```c
#include <avr/interrupt.h>
#include <avr/sleep.h>
int tick = 0; // Boolean to remember LED status
int repeat = 0; // Counter for number of repeats of timer
void setup() {
  cli(); // Clear global interrupt
  DDRB      ; // Set Pin 13 as output using bitmath
  DDRB      ; // Set Pin 10 as input using bitmath
  // Set control regs for Timer 2
  TCCR2A =        ; // Set to Mode Normal
  TCCR2B =        ; // Disable timer
  TIMSK2 =        ; // Turn ON timer overflow mask
  // Set control regs for PCINT interrupt
  PCICR |=        ; // Enable PCINT0 (Port B)
  PCMSK0 |=        ; // PCINT0 = Pin 10 of Digital port
  // Serial.begin(9600);
// For debug: Initialize serial monitor
  sei(); // Set global Interrupt
  // Use an appropriate sleep mode
  // Available Modes: SLEEP_MODE_X
  // where X can be IDLE/ADC/PWR_SAVE/STANDBY/EXT_STANDBY/PWR_DOWN
  set_sleep_mode(          );
}
// ISR for pin change interrupt capture
// when switch is pressed
ISR(PCINT0_vect) {
  // Serial.println("Switch press interrupt");
  // For debug
  PORTB       ; // Ensure the LED to OFF
  // Sample calc. to find # of repeats
  // and initial value for 2 sec
  // Clock cycle needed =  2*16M/1024 = 31,250
  // # of Repeat = 31,250/256 = 122.070312
  // After 122 repeats, remaining = 31,250-(256*122) = 18
  // Initial value = 256 - 18 = 238 = 0xEE
  TCNT2 =     ; // Set initial value for 2 sec count
  TCCR2B =        ; // Use pre-scale to 1024, start timer
}
// ISR for timer 2 when overflow (FF->00) occurs
ISR(TIMER2_OVF_vect) {
  // Serial.println("Timer 2 interrupt"); // For debug
  // After 2 sec, LED will be ON
  // After 3 sec, LED will be OFF
  if (tick == 0) { // LED is OFF
  repeat++;
  TCNT2 =    ; // Reset counter for next repeat
  if (repeat >    ) {
    PORTB       ; // Turn ON pin 13 after 2 sec
    // Like above, calculate # of repeats
    // and initial value for 3 sec
```

```
// Clock cycle needed =
// # of Repeat =
// After
// Initial value =
TCNT2 =        ; // Initial value for 3 sec count
tick = 1; // Set Boolean to represent LED status
repeat = 0; // Reset repeat counter
 }
}
if (tick == 1) { // LED is ON
 repeat++;
 TCNT2 =    ; // reset for next repeat
 if (repeat >   ) {
   PORTB            ; // Turn OFF pin 13 after 3 sec
   TCCR2B           ; // Disable timer
   tick = 0;  // Reset Boolean to represent LED status
   repeat = 0; // Reset repeat counter
   }
 }
}
void loop() {
 // Serial.println(tick);
// For debug: Display in serial monitor
 // Serial.println(repeat);
// For debug: Display in serial moni-tor
 sleep_mode(); // Do nothing, CPU to Sleep
}
```

Ultrasound Sensor with IIR Low Pass Filter with Arduino

In this assignment, we will investigate using use of IIR (Infinite Impulse Response) type Low Pass Filter (LPF) with an Arduino Uno. We will use ultrasound data as input to the filter. A test code of ultrasonic sensor is given at the end of this assignment (in Provided Code), and a sample code of IIR LPF is given in Chapter 4.

Hardware Setup:

Component required:

An ultrasonic sensor, a breadboard, and jumper wires.

Initial Test Procedure:

Setup an Arduino Uno board hardware in the following way with an ultrasonic sensor. The ultrasonic sensor has 4 connections: VCC, Trig, Edge, Gnd. Connect them as follows:

Ultrasonic sensor connection	Arduino Uno connection
VCC	5 V
Trig	Digital pin 7
Edge	Digital pin 6
Gnd	GND

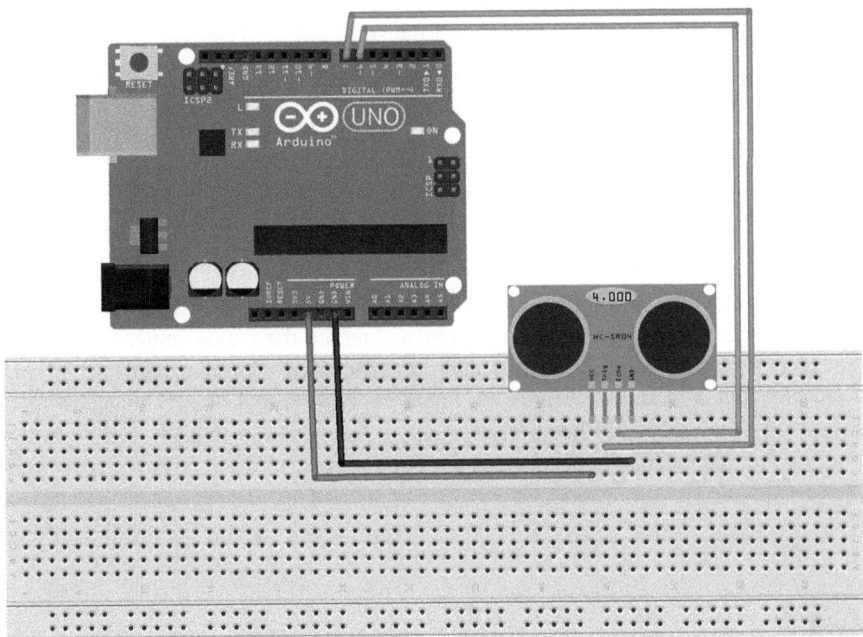

Copy the code from the Provided Code (below) to Arduino Sketch. Compile and download it on Arduino. Open "Serial Plotter" as the code is running. Now bring a solid object (e.g., Book) close to the front of ultrasonic sensor (~10 cm) and slowly move it away (~20 cm). Then suddenly move it away. If your setup is correct, your serial plotter should show something like this:

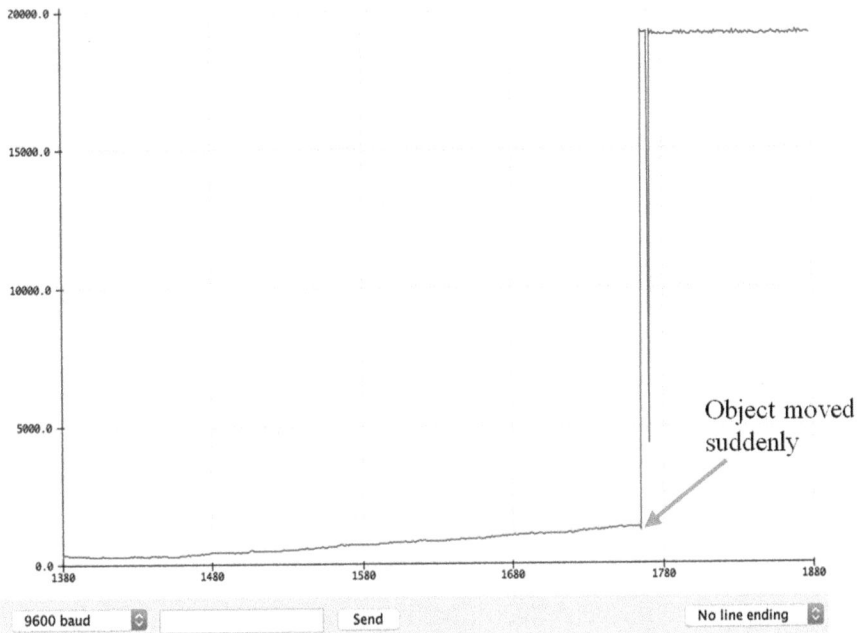

You can try other objects. Objects will flat solid face and compare with rounded object. Try moving the object slowly, quickly, or suddenly and observe how serial plotter data changes.

Task 1

Your first task in this assignment is to convert the y-axis data to distance (in cm). To do this, you can place your object at 10 cm, and record the y-axis value. Then move the object by 1 cm away (i.e., 11 cm) and record the value. Continue this until 30 cm. Then plot this data and determine the linear expression that closely matches your data. Now implement the linear expression in the code, so that the serial plotter y-axis value represents the distance of the object (in cm). *Note: To make your data even robust, you can take multiple measurements and use average statistics.*

Task 2

After you have modified the code to show data in distance (cm), try to resolve the sudden movement data issue. With the current setup, data can change suddenly (jitter) leading to problematic behavior of the actuator (e.g., motor of self-driving car). One of the approaches to deal with these jitters is to add a Low Pass Filter (LPF) block to smooth out the data. For this task, you need to write code of an IIR LPF filter which takes the distance data as input, and produces smoothed out data. *Note that the alpha parameter of the LPF code determines filter response. For smoother filter output response, a smaller alpha value is needed.*

Assignment report:

In the assignment report, provide the followings:

(a) Briefly summarize the work including your objective, procedure, and key results of the two tasks.
(b) Describe how you determined the linear expression in Task 1. Provide table of collected data and corresponding plot for the linear expression determination.
(c) Explain how you modified the IIR LPF code to perform Task 2.
(d) Provide the final completed code for both tasks with sufficient comments to ensure the code is understandable.
(e) Provide snapshots confirming successful download of the code from computer to Arduino board for both tasks.
(f) Provide snapshots of serial plotter showing (i) implemented provided code, (ii) after converting the y-axis data to distance (cm), and (iii) after implementing IIR LPF.
(g) Did you encounter any technical challenge? How did you resolve it?

Provided Code:

Code to test hardware setup:

```
const int pingPin = 7; // Trigger Pin of Ultrasonic Sensor
const int echoPin = 6; // Echo Pin of Ultrasonic Sensor
// Connect Ultrasonic sensor VCC to 5 V, and Gnd to 0 V
int duration;
void setup() {
  Serial.begin(9600); // Starting Serial Terminal
}
void loop() {
// Ultrasound sensor ping
  pinMode(pingPin, OUTPUT);
  digitalWrite(pingPin, LOW);
  delayMicroseconds(2);
  digitalWrite(pingPin, HIGH);
  delayMicroseconds(10);
  digitalWrite(pingPin, LOW);
// Ultrasound sensor echo catch
  pinMode(echoPin, INPUT);
  duration = pulseIn(echoPin, HIGH);
// Send raw data to Serial port
  Serial.println(duration);
// Wait before next ping
  delay(100);
}
```

Index

© Springer Nature Switzerland AG 2021
B. I. Morshed, *Embedded Systems – A Hardware-Software Co-Design Approach*,
https://doi.org/10.1007/978-3-030-66808-2